Sat[urn] HVAC S[ystems] Field Guide

Produced by John Krigger and Chris Dorsi

**Illustrated by John Krigger, Bob Starkey,
Steve Hogan, and Mike Kindsfater**
This edition compiled by Darrel Tenter

The *Saturn HVAC Field Guide* describes the procedures used to evaluate and service forced air heating and cooling systems.

The companion volume *Saturn Building Shell Field Guide* outlines procedures for insulating, air-sealing, and improving the shading of existing homes.

The *Saturn Hydronic Systems Field Guide* includes procedures for evaluating and servicing steam and hot-water space-heating systems.

The companion volume *Saturn Energy Auditor Field Guide* describes the procedures used to analyze the performance of existing homes.

In compiling this publication, the authors have benefited from the experience of many individuals who have reviewed our documents, related their experiences, or published information from which we've gained insight. Though we can't name everyone to whom we're indebted, we acknowledge the specific contributions of the following people: Martha Benewicz, Michael Blasnik, Anthony Cox, Bob Davis, Jim Davis, R.W. Davis, Rob de Kieffer, Rick Karg, Rudy Leatherman, Dave Like, Bruce Manclark, David Miller, Rich Moore, Gary Nelson, Judy Roberson, Russ Rudy, Russ Shaber, Cal Steiner, Ken Tohinaka, John Tooley, Bill Van Der Meer, and Doug Walter. We take full responsibility, however, for the content and use of this publication.

SATURN
RESOURCE MANAGEMENT

Foreword

The Saturn *HVAC Systems Field Guide* outlines specifications and procedures for improving the performance and efficiency of HVAC systems in residential and light-commercial buildings. This guide encourages technicians and contractors to use a building-science based approach to HVAC service.

We assume that the reader is experienced in working on HVAC systems. We don't cover every detail of every energy-conserving procedure in the interest of brevity and clear communication.

Chapter 1 outlines the whole-house approach we've found necessary to performing effective HVAC evaluation and service.

Chapter 2 addresses combustion safety, efficiency testing, and heating system service.

Chapter 3 provides guidelines for assessing chimney condition and function, and includes important tips for troubleshooting this sometimes problematical system.

Chapter 4 includes inspection and replacement standards for forced air combustion heating systems. It also includes protocols for wood stoves and electric heating systems.

Chapter 5 addresses the complicated subject of forced air delivery systems. Evaluating airflow and duct leakage is the major focus.

Chapter 6 focuses on the heat-exchange and efficiency aspects of heat pumps and air conditioners. Evaporative coolers are included.

Chapter 7 introduces a simple set of options for designing and sizing mechanical ventilation systems. If you've been mapping out the best way to protect the indoor air quality in the homes you work with, we think you'll appreciate the streamlined approach we've developed.

Chapter 8 provides guidelines for evaluating and replacement of various types of domestic hot water systems.

The Saturn *Field Guides* have benefitted greatly over the years from the generous feedback of our readers. Please help continue this process by sending us your comments and suggestions.

John Krigger
jkrigger@srmi.biz

Chris Dorsi
cdorsi@srmi.biz

TABLE OF CONTENTS

1: HVAC Energy Efficiency Service

2: Evaluating Combustion and Venting

3: Venting and Combustion Air

4: Heating System Installation

5: Evaluating Forced-Air System Performance

6: Cooling and Heat Pumps

7: Mechanical Ventilation

8: Water-Heating Energy Savings

Table of Contents

Chapter 1: HVAC Energy Efficiency Service

As an HVAC technician, you can improve customer satisfaction, out-perform your competition, and open new sales and marketing opportunities by specializing in energy efficiency.

This field guide outlines how to integrate energy-efficiency testing and improvements into your existing installation and service procedures.

Since most homes and small commercial buildings are heated and cooled by forced-air systems, this field guide concentrates on forced air.

1.1 Buildings Are Systems

A home's design should integrate the building shell with heating, cooling, and ventilation systems. A well-insulated and airtight home compliments a well-designed and installed HVAC system. And unfortunately, a drafty, unshaded, and poorly insulated home detracts from the performance of the HVAC system.

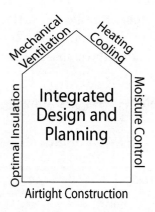

The building shell's poor thermal performance may overwhelm the HVAC system during extreme weather, causing customer complaints. Sometimes the best solution to a customer complaint involves adding insulation, sealing air leaks, or shading windows, rather than by trying to squeeze better performance out of the heating or cooling system.

You may have been blamed more than once for a poorly performing building shell. We suggest that you become better educated on how the home functions as a system. In particular, moisture problems are a frequent cause of respiratory problems and poor cooling performance. Many HVAC contractors are now working with insulation contractors and other trades to provide a more integrated energy-and-comfort packages for their customers.

1.2 EVALUATING THE BUILDING SHELL

The characteristics of the building shell determine the heating and cooling load of the building. Therefore, it's important to understand the basics of winter and summer heat flow through buildings. Winter heat loss is driven by the temperature difference between indoors and outdoors. Summer heat gain is driven primarily by solar heat gain, so shading and reflectivity are as important as insulation for controlling cooling costs.

A building's *thermal boundary* is a line where the insulation together with the air barrier are installed. Shading and reflectivity are also important to controlling the amount of solar heat that flows through the thermal boundary, particularly in the cooling season.

Table 1-1: Factors Affecting Heating and Cooling Load

Space-Conditioning Factors	% of heating	% of cooling
Temperature Difference	60–80%	5–15%
Air Exchange	20–40%	10–20%
Solar Heat Gain	*	40–60%
Internal Gains	*	10–20%
* Has a small heating effect, which is negligible in most homes but significant in well insulated and airtight homes and homes in warm climates.		

1.2.1 Insulation

America's building industry has placed too little emphasis on the energy efficiency of the building shell. Insulation is the most valuable investment a building owner can make in constructing or renovating a building. Attic or roof-cavity insulation is extremely important in all climates to limit heat loss and heat gain through the vulnerable roof area. Wall insulation is also very important. Both attic and wall insulation are often inadequate for wise energy use because of installation flaws and insufficient insulation thickness. Floor or foundation insulation is simply neglected in most homes, causing excessive energy waste especially in cool climates.

The building shell is supported by wood, metal and other non-insulating materials. The building shell's thermal resistance then depends on an average of the insulation and structural materials, which is usually less than the nominal value of the installed insulation (called: whole-wall R-value). For example, 12–22 percent of the typical surface of a framed wall is lumber and only 78–88 percent is insulated cavity. In the case of a 2-by-4 wall with R-11 insulation, the whole-wall R-value, accounting for this mix of materials on the surface of the wall is actually a little less than R-11. The whole-wall R-value can be reduced another 10–25 percent or more by installation flaws.

Thermal bridging describes the thermal-resistance effect of non-insulating materials like wood and metal at the thermal boundary. Steel has a thermal conductance 200 times that of wood and aluminum has a thermal conductance of 900 times that of wood. These materials can have a dramatic effect on a building component's R-value. Insulated wall sheathing should be standard practice to reduce thermal bridging through wood and metal studs.

Convection is air movement within or around the insulation, which reduces its thermal resistance. Convection is a particular problem with walls because they are vertical like chimneys. Any voids or small channels, created during installation, encourage convection in wall cavities. This convection carries heat through or around the insulation. Careful cutting and fitting of batts and careful dense-packing of loose-fill insulation minimizes convection.

1.2.2 Air Leakage and Ventilation

Blower door: This measuring device consists of a fan, mounted in a panel connected to manometers that read house pressure and airflow.

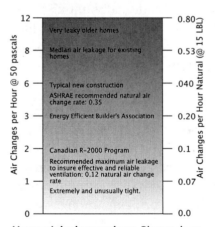

Home air leakage values: Blower door testing reveals a spectrum of different values, depending on construction characteristics.

Most buildings depend on air leakage to provide ventilation to occupants. The American Society of Heating, Refrigeration, and Air-Conditioning Engineers (ASHRAE) and the EPA recommend ventilation of 0.35 air changes per hour for acceptable indoor air quality. Most existing buildings fall with a range of 0.2 to 0.8 air changes per hour (ACHn). This range demonstrates the overall lack of planning to control air leakage in buildings.

An air-leakage rate of more than 0.8 is rare except in older housing, although modern homes with complex shapes, fireplaces, and dozens of recessed light fixtures may reach the upper parts of this air-leakage range.

In the past 20 years, building analysts have used blower doors to measure air leakage and to guide air-sealing work. A blower door is a fan installed in a doorway, equipped with airflow-measuring capability. Air sealing shouldn't happen without blower door testing because first you don't know whether air sealing is needed and second you can't evaluate the ventilation issues associated with air sealing.

Options for Air-Sealing and Ventilation

There are three common options for air sealing and ventilation.

1. Reduce air leakage by sealing large hidden air leaks with the goal of reducing air leakage to around 0.35 ACHn or some higher value depending on cost and feasibility. This is the most common approach. Ventilation is achieved with exhaust fans in the kitchen and bathroom. *See "Spot Ventilation" on page 176.*

2. Reduce air leakage to between 0.35 and 0.10 and install exhaust or supply whole-house ventilation. *See "Exhaust Ventilation" on page 182. See also "Supply Ventilation" on page 184.*

3. Reduce air leakage to 0.1 ACHn or less and install whole-house mechanical ventilation. This approach is usually part of a goal to reduce heating and cooling costs to a practical minimum level. *See "Balanced Ventilation Systems" on page 185.*

When choosing the third option, consider the building a part of the ducted air delivery system. Air leakage, more than 0.1 ACHn, interferes with the proper delivery of air to rooms because the ducted airflow rate is small. Any exchange of air through the building shell significantly affects the performance of a balanced whole-house ventilation system.

When you install a whole-house ventilation fan in a home with air leakage between 0.35 and 0.10 (option 2), the fan will add only about half of its rated airflow in CFM to the natural infiltration rate. This half-fan rule must be accounted for in sizing a continuously operating fan or determining its cycling schedule to ensure adequate ventilation air.

1.2.3 Shading and Reflectivity

Shading and reflectivity are very important during the cooling season, especially at the roof and windows. Despite the fact that reflective white roofs improve comfort and save 10–20 percent on cooling, most architects, builders, and homeowners select dark-colored roofing. Colored roofing materials typically absorb more than 75 percent of incident solar energy, while reflective roofing absorbs less than 25 percent.

Windows solar gain can be up to 40 percent of the accumulated heat removed by the air-conditioning system. Single-pane glass has a solar heat gain coefficient (SHGC) of 0.87, meaning that it transmits 87 percent of solar heat. The best spectrally selective insulated glass unit has a SHGC of less than 0.30.

Shading coefficient compares the solar transmittance of a glass assembly, with its window treatments, to single-pane glass. Many window-treatment manufacturers, offering for example window films and solar screens, list this shading coefficient. The shading coefficient is always 1 or less and always greater than the SHGC.

1.3 Evaluating HVAC for Energy Efficiency

This field guide contains standards, specifications, and procedures for evaluating HVAC systems and servicing them for energy-efficiency. The authors assume that the reader is familiar with HVAC operation. Understanding the operation of an HVAC system is essential to improving its efficiency and requires the following specific knowledge.

- Understanding of combustion systems, including heat exchangers, burners, and venting systems.
- Understanding of the refrigeration cycle.

- Understand electricity, simple electrical devices, and electric circuits used in HVAC systems.

- Ability to use common testing equipment including combustion analyzers and digital multimeters.

1.3.1 Summary of HVAC Field Studies

Efficiency, comfort, and safety problems with residential heating and cooling systems are rampant throughout the U.S. The main cause of these problems is a failure to measure performance during installation and service. An EPA-sponsored report titled, *National Energy Savings Potential from Addressing Residential HVAC Installation Problems* reviewed many individual studies of installation flaws, and provided the following summary.

Table 1-2:   Summary of HVAC Performance Field Studies

Installation-Related Problem	%*	Savings Potential
Duct air leakage (Ave. 270 CFM_{25})	70%	17% Ave.
Inadequate airflow	70%	7% Ave.
Incorrect charge	74%	12% Ave.
Oversized by 50% or more	47%	2–10%

* Percent of tested homes found with a significant problem.
* The number of homes of the duct-leakage studies was around 14,000; the number for the other problems was over 400 each.
Report sponsored by Environmental Protection Agency (EPA)

1.3.2 Commissioning HVAC Systems

While many customers may still want the minimum acceptable installation and service work, many others are willing to pay more for optimized installation and service.

We suggest using measurements to verify safe and efficient system performance. Engineers call this process *commissioning*.

Commissioning means testing the HVAC system to determine whether it is performing safely and efficiently and then making necessary adjustments to optimize efficiency and performance.

Commissioning a residential gas furnace involves, at minimum, the following steps.

- ✓ Performing combustion test to measure oxygen, temperature, draft, and carbon monoxide in the flue gases.

- ✓ Estimating airflow to verify conformance with manufacturer's and designer's specifications.

- ✓ Measuring house pressure in the combustion zone to ensure that negative pressure, if present, is within acceptable limits.

- ✓ Measuring input and gas pressure, if necessary to troubleshoot combustion problems.

- ✓ Measuring duct leakage and taking steps to reduce the duct air leakage to a specific standard.

Commissioning a residential air conditioner or heat pump involves the following steps.

- ✓ Measuring airflow and taking whatever steps necessary to improve airflow, including adding ducts, enlarging ducts, replacing filters, cleaning the blower, cleaning indoor coil, and balancing room airflows.

- ✓ Measuring one of two indicators of refrigerant charge and adding or removing refrigerant as necessary to create the correct charge.

✓Verifying correct functioning and settings of electric controls including fan speeds for heating and cooling, outdoor thermostat for heat pumps, thermostat functioning, and appropriate settings for programmable thermostats.

✓Noting whether excessive air leakage, moisture problems, lack of insulation, or excessive solar gain through windows is overpowering the cooling system or causing high air-conditioning costs.

✓Measuring duct leakage and taking steps to reduce the duct air leakage to a specific standard.

Access for Installation and Service

Technicians frequently are provided with too little space to install the equipment with adequate clearance for maintenance. If technicians don't have easy access to equipment, which requires periodic inspection and cleaning, service may be neglected. The HVAC contractor should be part of the design team so that he can ensure there is ample space reserved for the equipment and ductwork.

1.3.3 Equipment Sizing and Selection

Equipment selection should be based on calculated heating and cooling loads and calculated duct sizes. Since hand calculations are time-consuming and tedious, progressive contractors use computer software for these design calculations.

Air handlers with their heating and cooling equipment tend to be significantly oversized. One reason is that HVAC contractors don't have adequate information or confidence about the building where they are installing the system. For example, is the natural air-leakage rate 0.1 or 1 air changes per hour? Without a blower door test, how is the HVAC contractor to know?

The Manual J computer software contains a safety factor of around 25%. If the contractor adds another 25% safety factor that makes 50% to 60% oversizing. If the wholesaler suggests bumping the output up another notch, the output can end up double what the home actually requires during outdoor design conditions.

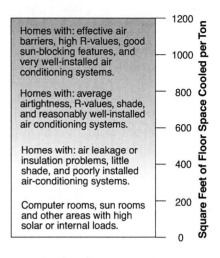

Unfortunately, these excessive safety factors may actually be needed to counteract installation problems, which sap the air handler's heating and cooling capacity, including the following.

Existing homes and their cooling loads: The characteristics of the building shell and climate, along with the quality of air-conditioner installation, determine how many square feet of floor space can be cooled with each ton of air-conditioner capacity.

- Duct air leakage.

- Inadequate airflow.

- Incorrect refrigerant charge.

Preventing or repairing these installation problems, using this field guide, eliminates the need for compensatory oversizing. Better planning and cooperation between general contractor and HVAC contractor can allow the HVAC contractor more space and time to provide a quality installation. And, a high-quality insulation package and an effective air barrier can allow the HVAC contractor to install the minimum size air handler with minimum heating and cooling capacity, which provides the following benefits, compared to current typical design and installation practices.

1. Smaller ducts occupy less space and are less likely to be deformed when installed in tight building cavities.

2. A lower designed airflow is easier to achieve.

3. The system operates at higher efficiency and lower operating costs.

4. The system operates more predictably and with less problems.

5. The equipment lasts longer.

6. The initial cost is lower.

7. Comfort is better and operation is quieter.

Systems with two-stage compressors and two-stage gas furnaces can aide in achieving some of these goals when the HVAC contractor can't affect the home's design and construction. The first stage of heating or cooling is a far better match for most conditions than the unit's maximum output.

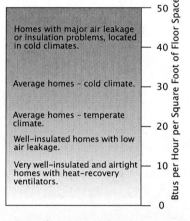

Existing homes and their heating loads: The characteristics of the building shell and climate determine how many BTUs/hour of heat are needed per square foot of floor space at design winter conditions.

The best homes being built today function as integrated systems. The building shell minimizes heat loss in winter and heat gain in summer, and such a home requires a small heating and cooling system. Much of the year, neither heating nor cooling is needed. Since the shell is airtight, mechanical ventilation systems are included in these exemplary homes.

Customer comfort and satisfaction with these exemplary homes are very high. Some home retrofitters are achieving a similar high standard for the building shell and HVAC system. Considering the improvements in comfort, operating costs, health, and safety that are possible from higher-than-normal standards, the additional effort and initial expense is an excellent investment.

HVAC Energy Efficiency Service

CHAPTER 2:   EVALUATING COMBUSTION AND VENTING

The effective performance of burners and their venting systems is essential for safe and efficient combustion. The burners must not produce excessive carbon monoxide (CO), which is a poison. The venting system must convey the combustion gases outside the building.

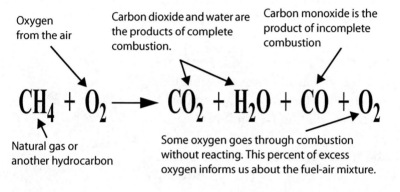

Oxygen from the air

Carbon dioxide and water are the products of complete combustion.

Carbon monoxide is the product of incomplete combustion

$$CH_4 + O_2 \longrightarrow CO_2 + H_2O + CO + O_2$$

Natural gas or another hydrocarbon

Some oxygen goes through combustion without reacting. This percent of excess oxygen informs us about the fuel-air mixture.

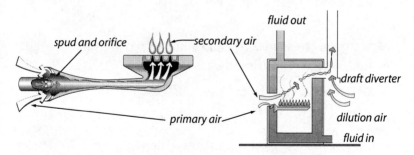

fluid out

spud and orifice

secondary air

draft diverter

primary air

dilution air

fluid in

Atmospheric, open-combustion gas burners: Combustion air comes from indoors in open-combustion appliances. These burners use the heat of the flame to pull combustion air into the burner. Dilution air, entering at the draft diverter, prevents over-fire draft from becoming excessive.

2.1 ESSENTIAL COMBUSTION SAFETY TESTS

The Building Performance Institute (BPI) requires that essential combustion safety tests be performed as part of all energy conservation jobs. BPI requires natural-gas leak-testing and CO testing for all appliances. For naturally drafting appliances, either a worst-case venting test or zone-isolation test is also necessary.

BPI considers naturally drafting appliances and venting systems to be obsolete for both efficiency and safety reasons. BPI strongly recommends that these obsolete appliances be replaced with modern direct-vent or power-vent combustion appliances.

2.1.1 Leak-Testing Gas Piping

Natural gas and propane piping systems may leak at their joints and valves. Find gas leaks with an electronic combustible-gas detector, often called a gas sniffer. A gas sniffer will find all significant gas leaks if used carefully. Remember that natural gas rises from a leak and propane falls, so position the sensor accordingly.

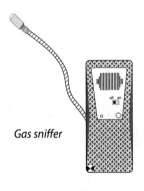

Gas sniffer

- ✓ Sniff all valves and joints with the gas sniffer.

- ✓ Accurately locate leaks using a non-corrosive bubbling liquid, designed for finding gas leaks.

- ✓ All gas leaks must be repaired.

- ✓ Replace kinked or corroded flexible gas connectors.

- ✓ Replace flexible gas lines manufactured before 1973. The date is stamped on a date ring attached to the flexible gas line.

2.1.2 Carbon Monoxide (CO) Testing

CO testing is essential for evaluating combustion and venting. Measure CO in the vent of every combustion appliance you inspect and service. Measure CO in ambient air.

Vent Testing for CO

Testing for CO in the appliance vent is a part of combustion that takes place under worst-case conditions. If CO is present in undiluted combustion byproducts more than 100 parts per million (ppm), the appliance fails the CO test.

Ambient Air Monitoring for CO

BPI standards require technicians to monitor CO during testing to ensure that air in the combustion appliance zone (CAZ) doesn't exceed 35 parts per million. If ambient CO levels in the combustion zone exceed 35 parts per million (ppm), stop testing for the your own safety. Ventilate the CAZ thoroughly before resuming combustion testing. Investigate indoor CO levels of 9 ppm or greater to determine their cause.

Table 2-1: Testing Requirements for Combustion Appliances and Their Venting Systems

Appliance/Venting System	Required Testing
All direct-vent or power-vent combustion appliances	Gas leak test CO test at flue-gas exhaust outdoors Confirm venting system connected
Combustion appliances (with naturally drafting chimneys) in a mechanical room or attached garage supplied with outdoor combustion air	Gas leak test CO test Confirm that CAZ is disconnected from main zone
Naturally drafting chimney and appliance located within home	Gas leak test CO test Venting inspection Worst-case draft and depressurization testing

2.1.3 Worst-Case Testing for Atmospheric Venting Systems

Depressurization is the leading cause of backdrafting and flame roll-out in furnaces and water heaters that vent into naturally drafting chimneys. The best option is to replace the older appliances and their naturally drafting venting systems with direct-vented or power-vented appliances with airtight venting systems. However, if the atmospheric appliances and venting systems must remain, perform the worst-case testing procedures documented here.

Worst-case vent testing uses the home's exhaust fans, air handler, and chimneys to create worst-case depressurization in the combustion-appliance zone (CAZ). The CAZ is an area containing one or more combustion appliances. During this worst-case testing, you can test for spillage, measure the indoor-outdoor pressure difference, and measure CO level.

Worst-case conditions do occur, and venting systems must exhaust combustion byproducts even under these extreme conditions. Worst-case vent testing exposes whether or not the venting system exhausts the combustion gases when the combustion-zone pressure is as negative as you can make it. A sensitive digital manometer is the best tool for accurate measurements of both combustion-zone depressurization and chimney draft.

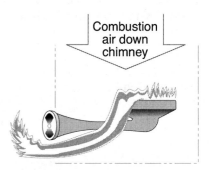

Flame roll-out: Flame roll-out, a serious fire hazard, can occur when the chimney is blocked, the combustion zone is depressurized, or during very cold weather.

Take all necessary steps to reduce spillage and strengthen draft as necessary based on testing.

Worst-case depressurization: Worst-case testing is used to identify problems that weaken draft and restrict combustion air. The testing described here is intended to isolate the negative-pressure source.

A reading more negative than –5 pascals indicates a significant possibility of backdrafting.

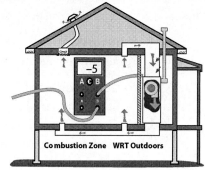

Combustion Zone   WRT Outdoors

Table 2-2:   Maximum CAZ or Mechanical Room Depressurization for Various Appliances

Appliance	Maximum Depressurization
Direct-vent appliance	50 pa (0.20 IWC)
Pellet stove with draft fan and sealed vent	15 pa (0.06 IWC)
Atmospherically vented oil and gas systems	5 pa (0.02 IWC)
Oil power burner and fan-assisted (induced-draft) gas*	
Closed controlled combustion	
Decorative wood-burning appliances	
Atmospherically vented water heater	2 pa (0.008 IWC)

*Individual fan-assisted (induced-draft) appliances with no vent hood attached to intact B-vent and oil appliances with flame retention head power burners are likely to vent safely at greater than 5 pascals depressurization but not enough test data is available to set a higher limit at this time. Since the appliances are possibly connected to an unsealed chimney and most spillage is through joints and the barometric damper these systems are included in the 5pa limit.

Worst-Case Depressurization, Spillage, and CO

Start with all exterior doors, windows, and fireplace damper(s) closed and measure the base pressure.

1. Set all combustion appliances to the pilot setting or turn them off at the service disconnect.

2. Measure and record the base pressure of the combustion appliance zone (CAZ) with reference to outdoors. If the digital manometer has a self-zeroing or "base" function, use this zeroing function now.

Next, establish worst-case conditions and measure the maximum worst-case depressurization.

1. Turn on the dryer and all exhaust fans.

2. Close interior doors that make the CAZ pressure more negative.

3. Turn on the air handler, if present, and leave on if the pressure in the CAZ becomes more negative.

4. Measure the net change in pressure from the CAZ to outside, correcting for the base pressure previously. Record the "worst-case depressurization" and compare to the table entitled, *"Maximum CAZ or Mechanical Room Depressurization for Various Appliances"* on *page 29* for the tested appliance.

Negative Versus Positive Draft:
With positive draft air flows down the chimney and out the draft diverter. A smoke bottle helps distinguish between positive and negative draft in atmospheric chimneys.

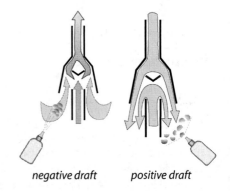

negative draft   *positive draft*

Finally, fire the combustion appliances and test for spillage and CO.

1. Fire the appliance with the smallest BTU capacity first and then the next largest and so on.

2. Test for spillage at the draft diverter with a smoke generator, a lit match, or a mirror. Note whether combustion byproducts spill and how long after ignition that the spillage stops.

3. Test CO in the undiluted flue gases at 5 minutes.

4. If spillage in one or more appliances continues under worst-case for 1 minute or more, test the appliance again under natural conditions and consider the improvements below to eliminate spillage.

2.1.4 Measuring Chimney Draft

If the presence or absence of spillage doesn't provide enough diagnostic information to troubleshoot a venting problem, measure the chimney draft.

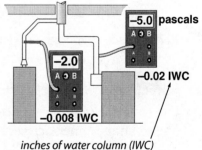

inches of water column (IWC)

Measuring draft: Measure chimney draft *downstream* of the draft diverter.

1. Drill a hole in the vent connector a foot or two downstream of the draft diverter or draft inducer (in the case of a draft-induced 80+% furnace).

2. Measure draft from startup until blower activation.

3. Note changes related to wind, blower activation, opening a window, or activating exhaust appliances.

2.1.5 Eliminating Spillage and Improving Insufficient Draft

When you observe spillage or measure weak draft, investigate the cause. Open a window or door to observe whether the addition of combustion air improves draft. If so, the problem usually is depressurization. If opening a window has no effect on spillage or draft, inspect the venting system for blockage, leaks, or poor design.

Chimney Improvements to Eliminate Spillage and Improve Draft

Chimneys are a frequent cause of backdrafting, spillage, and weak draft. Consider the following for improving chimney performance.

- ✓Remove chimney obstructions.

- ✓Remove single-wall vent connectors and replace with double-wall Type B or Type L vent.

- ✓Repair disconnections or leaks at joints and where the vent connector joins a masonry chimney.

- ✓Increase the pitch of horizontal sections of vent.

- ✓If the masonry chimney is deteriorated, consider installing a new chimney liner. *See page 70.*

- ✓Measure the size of the vent connector and chimney and compare to vent-sizing information listed in Chapter 13 of the *National Fuel Gas Code (NFPA 54).* A vent connector or chimney liner that is either too large or too small can cause weak draft.

- ✓Increase the height of the chimney.

- ✓If wind is causing erratic draft, consider installing a wind-dampening chimney cap.

Table 2-3:   Draft Problems and Solutions

Problem	Possible Solutions
Adequate draft never established	Remove chimney blockage, seal chimney air leaks, or provide additional combustion air as necessary.
Blower activation weakens draft	Seal leaks in the furnace and in nearby return ducts. Isolate the furnace from nearby return registers.
Exhaust fans weaken draft	Provide make-up or combustion air if opening a door or window to outdoors strengthens draft during testing.
Closing interior doors during blower operation weakens draft	Add return ducts, grills between rooms, or jumper ducts.

Duct Improvements to Solve Draft Problems

Return-duct leakage or unbalanced ducted airflow can depressurize a CAZ and cause spillage. Consider the following remedies.

✓ Repair all return-duct leaks near furnace.

✓ Isolate furnace from return registers by air-sealing.

✓ Improve balance between supply and return air by installing new return ducts, transfer grills, or jumper ducts. *See page 126.*

Reducing Depressurization from Exhaust Appliances

Consider the following remedies to depressurization caused by the home's exhaust appliances.

✓ Isolate furnace from exhaust fans and clothes dryers by air-sealing between the CAZ and zones containing these exhaust devices.

✓ Reduce capacity of large exhaust fans.

✓Provide make-up air for dryers and exhaust fans and/or provide combustion-air inlet(s) to combustion zone. *See page 77.*

2.1.6 Zone Isolation Testing for Atmospherically Vented Appliances

An isolated CAZ improves the safety of atmospherically vented appliances. The CAZ is isolated if it obtains combustion air only from outdoors. An isolated CAZ doesn't require worst-case depressurization and spillage testing. However the zone must be visually inspected for connections with the home's main zone and tested for isolation.

1. Look for connections between the isolated CAZ and the home. Examples include joist spaces, transfer grills, leaky doors, and holes for ducts or pipes.

2. Measure a base pressure from the CAZ to outdoors.

3. Perform 50-pascal blower door depressurization test. The CAZ-to-outdoors pressure should not change more than 5 pascals during the blower door test.

4. If the CAZ-to-outdoors pressure changed more than 5 pascals, perform air sealing to completely isolate the zone and retest as described above. Or alternatively perform a worst case depressurization and spillage test as described in *"Worst-Case Depressurization, Spillage, and CO" on page 30.*

2.2 Combustion Testing Gas Furnaces

The goal of a combustion analysis is to quickly analyze combustion and heat exchange. Within ten minutes of activating a burner, you can know its most critical operating parameters. This information saves time and informs your decision making.

2.2.1 Combustion Efficiency Testing and Adjustment

Modern flue-gas analyzers measure O_2, CO, and flue-gas temperature. The better models also measure draft. Flue-gas analyzers also calculate combustion efficiency or steady-state efficiency (SSE). SSE is calculated from O_2 and flue-gas temperature. The SSE is usually 1-3 percentage points higher than the Annual Fuel Utilization Efficiency (AFUE). Using a modern electronic flue-gas analyzer, an experienced technician can perform a combustion test quickly.

Flue-gas temperature is an important indicator of heat exchange. A low fuel-gas temperature is usually an indicator of efficient performance. However, if the flue-gas temperature is too low, acidic condensation forms in the vent. This acidic condensate can rust metal vents and deteriorate masonry chimneys.

A common furnace-efficiency problem is incorrect fuel input, and high or low O_2. Optimizing the steady-state efficiency (SSE) and fuel-air mixture, using combustion testing, can save 2–8% of the fuel consumption.

Perform the following procedures to verify a combustion system's correct operation.

✓Perform a combustion test using a electronic flue-gas analyzer. Note CO, O_2, and flue-gas temperature. CO should be less than 100 ppm. O_2 should be between 6% and 9%. Recommended flue gas temperature depends on the type of heating system and is listed in the table titled, *"Combustion Standards for Gas Furnaces" on page 39.*

70+ Furnace: Sample flue gases in the draft diverter. Measure draft in the vent connector above the furnace.

80+ Furnace: Measure draft and sample flue gases in the vent connector above the furnace.

✓Measure temperature rise (supply minus return temperatures). Temperature rise should be within the manufacturer's specifications. Estimate the airflow from the furnace's blower specifications and compare the furnace's rated output to output estimated by the table titled: *"Gas-Furnace Output from Temperature Rise and Airflow (kBTUH)" on page 38*

✓If O_2 is low, or the estimated output from the table is high, decrease gas pressure to bring O_2 to 6% or slightly above. Low O2 can be a sign of inadequate combustion air or excessive fuel input. Excessive CO levels might result from this condition.

✓If O_2 is high, or the estimated output from the table is low, increase gas pressure to 6% O_2 if possible as long as you don't create CO.

✓Increasing gas pressure may increase temperature rise and flue-gas temperature. Before increasing gas pressure, you should believe that you can increase airflow to compensate. Otherwise you may not achieve any energy savings.

Troubleshooting Temperature Rise

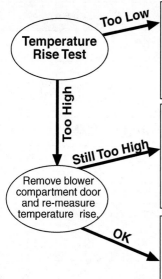

Temperature Rise Test

Too Low

Too High

Still Too High

OK

Remove blower compartment door and re-measure temperature rise.

Temperature Rise is Too Low
1. Look for signs of corrosion in the vent and heat exchanger.
2. Test gas input, and increase if too low.
3. Check for return air leakage from outdoors.
4. Reduce fan speed.

Supply airflow is inadequate
1. Clean blower. Increase blower speed.
2. Find and remove restrictions in the supply ducts and registers.
3. Add additional supply branches to hard-to-heat areas.
4. Increase size of supply ducts and registers to hard-to-heat areas.

Return airflow is inadequate
1. Look for restrictions in the return ducts and registers.
2. Clean or replace filter
3. Clean blower. Increase blower speed.
4. Test gas input and reduce if too high.
5. Clean or remove AC coil.
6. Install new return air duct or jumper duct.
7. Install turning vanes in 90° main return

Table 2-4: Gas-Furnace Output from Temperature Rise and Airflow (kBTUH)

CFM	Temperature Rise (Supply F° – Return F°)					
	30	40	50	60	70	80
600	19	26	32	39	45	52
700	23	30	38	45	53	61
800	26	35	43	52	60	69
900	29	39	49	58	68	78
1000	32	43	54	65	76	86
1100	36	48	59	71	83	95
1200	39	52	65	78	91	104
1300	42	56	70	84	98	112
1400	45	61	76	91	106	121
1500	49	65	81	97	113	130
1600	52	69	86	104	121	138
1700	55	73	92	110	129	147
1800	58	78	97	117	136	156
1900	62	82	103	123	144	164
2000	65	86	108	130	151	173
Output kBTUH= 1.08 CFM x ΔT						

If you know the airflow through the furnace from measurements described in *"Evaluating Forced-Air-System Airflow" on page 102*, you can use the table above to check whether output is approximately what the manufacturer intended. Dividing this output by measured input from *"Measuring BTUH input on Natural Gas Appliances" on page 45* gives you another check on the steady-state efficiency.

Table 2-5:   Combustion Standards for Gas Furnaces

Performance Indicator	SSE 70+	SSE 80+	SSE 90+
Combustion-zone pressure (Pa)	–4 Pa.	–5 Pa.	–10 Pa.
Carbon monoxide (CO) (ppm)	≤ 100 ppm	≤ 100 ppm	≤ 100 ppm
Stack temperature (°F)	350°–475°	325°–450°	≤ 120°
Temperature rise (°F)	40–70°*	40–70°*	30–70°*
Oxygen (%O2)	5–10%	4–9%	4–9%
Gas pressure Inches (IWC)	3.2–4.2 IWC*	3.2–4.2 IWC*	3.2–4.2 IWC*
Steady-state efficiency (SSE) (%)	72–78%	78–82%	92–97%
Draft (Pa)	–5 Pa	–5 Pa	+25–60 Pa

* pmi = per manufacturer's instructions

Drilling and Patching Vents for Combustion Testing

For single-wall metal vents, drill a quarter-inch hole and patch it with high-temperature silicone caulking.

For double-wall metal vent, drill through both layers and install a $5/16$-inch automotive lag screw through the two-layer hole. Seal the lag screw's cap with high-temperature silicone.

For plastic vents, drill in a vertical section and drill slightly downwards. This prevents condensate from flowing into the hole. Again, seal the hole with high-temperature silicone.

Table 2-6: Carbon Monoxide Causes and Solutions

Cause	Analysis & Solution
Flame smothered by combustion gases.	Chimney backdrafting from CAZ depressurization or chimney blockage.
Burner or pilot flame impinges.	Align burner or pilot burner. Reduce gas pressure if excessive.
Inadequate combustion air with too rich fuel-air mixture.	O_2 is $\leq 6\%$. Gas input is excessive or combustion air is lacking. Reduce gas or add combustion air.
Blower interferes with flame.	Inspect heat exchanger. Replace furnace or heat exchanger.
Primary air shutter closed.	Open primary air shutter.
Dirt and debris on burner.	Clean burners.
Excessive combustion air cooling flame.	O_2 is $\geq 11\%$. Increase gas pressure.

2.2.2 Critical Furnace-Testing Parameters

The following group of furnace-testing parameters are the most important ones because they tell you how efficient and safe the furnace currently is and how much you might be able to improve efficiency. Learn to use these measurements to analyze the combustion process.

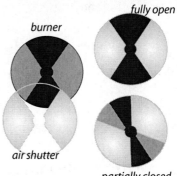

Carbon monoxide (CO) (ppm): Poisonous gas indicates incomplete combustion. Adjusting combustion to produce less than 100 ppm is almost always possible with gas-pressure adjustments, primary-air adjustments, or burner maintenance.

Primary air adjustment: Primary air shutters are usually fully open for natural gas combustion and partially closed for propane depending on flame characteristics. Too much primary air can cause noise and flame lift-off. Too little causes a lazy flame, searching for air. A good flame is hard and blue with an inner and outer mantle.

Oxygen (percent): Indicates the percent of excess air and whether fuel-air mixture is correct. Efficiency increases as oxygen decreases because excess air carries heat up the chimney. Percent O_2 may also indicate the cause of CO as either too little of too much combustion air.

Flue-gas temperature: The critical heat-exchange measurement: Flue-gas temperature is directly related to furnace efficiency. Too high flue-gas temperature wastes energy and too-low flue-gas temperature causes condensation.

Airflow: The furnace airflow is important to evaluate and is related to both flue-gas temperature and temperature rise. During testing, you can increase airflow and see the effect on flue-gas temperature and temperature rise. Increasing airflow can increase efficiency especially after increasing gas pressure to reduce the oxygen level in the combustion gases without increasing flue-gas temperature.

Duct leakage to outdoors: By pressurizing both the house and ducts, you can measure how much ducted air is leaking to outdoors. This value is directly related to the energy savings you can expect from duct sealing.

2.2.3 Other Relevant Furnace-Testing Parameters

The following furnace-testing parameters are interesting and important but less critical than those listed in the proceeding section. Learn to use these measurements, when the major indicators aren't enough to provide the information you need to thoroughly understand combustion and heat exchange.

Temperature rise: Manufacturers specify an acceptable range. A clue about heat exchange, airflow, firing rate, and efficiency.

Gas input: Indicates whether the appliance has the correct firing rate, specified by the manufacturer.

Gas pressure: Measured with a gas-pressure manometer, connected to the gas valve or manifold. Adjust within the range of pressures shown in *Table 2-5 on page 39*.

Steady-state efficiency (percent): Verifies that the SSE is within the correct range. An 80+ furnace functioning in the high 80 range may condense, corroding its vent. A 90+ condensing furnace, functioning in the high 80 range, may not be condensing like it should.

Evaluating Combustion Air

Inadequate combustion air will show itself through a low O_2 reading. If the O_2 reading is ≥6% with CO at ≤100 ppm, you can assume that there is adequate combustion air. O_2 at ≥6% with CO at ≤100 ppm indicates that there is no need for an outdoor combustion air duct or de-rating of a combustion furnace for altitude.

2.3 Prescriptive Inspection of Gas Combustion Systems

Gas burners should be inspected and maintained every 2 to 4 years. These following specifications apply to gas furnaces, water heaters, and space heaters.

2.3.1 Inspection

Perform the following inspection procedures and maintenance practices on all gas-fired furnaces, water heaters, and space heaters, as necessary. The goal of these measures is to reduce carbon monoxide (CO), stabilize flame, and verify the operation of safety controls.

✓ Look for soot, melted wire insulation, and rust in the burner and manifold area outside the fire box. These signs indicate flame roll-out, combustion gas spillage, and CO production.

✓ Inspect the burners for dust, debris, misalignment, flame-impingement, and other flame-interference problems. Clean, vacuum, and adjust as needed.

✓ Inspect the heat exchanger for leaks. *See page 135.*

✓ Furnaces should have dedicated circuits with fused disconnects. Assure that all 120-volt wiring connections are enclosed in covered electrical boxes.

✓ Determine that the pilot is burning (if equipped) and that main burner ignition is satisfactory.

✓ Test pilot-safety control for complete gas valve shutoff when pilot is extinguished.

✓ Check the thermostat's heat-anticipator setting. The thermostat's heat-anticipator setting should match the measured current in the 24-volt control circuit.

✓ Check venting system for proper size and pitch. *See page 63.*

✓ Check venting system for obstructions, blockages, or leaks.

✓ Test to ensure that the high-limit control extinguishes the burner before furnace temperature rises to 200° F.

✓ Measure gas input, and observe flame characteristics if soot, CO, or other combustion problems are present.

✓ Open a window while testing for CO to see if CO is reduced by increasing combustion air. *See page 77.*

2.3.2 Maintenance and Adjustment

Proceed with burner maintenance and adjustment when:

✓ CO is greater than 100 ppm.

✓ Visual indicators of soot or flame roll-out exist.

✓ Burners are visibly dirty.

✓ Measured draft is inadequate.

✓ The appliance has not been serviced for two years or more.

Gas-burner maintenance includes the following measures.

✓ Remove causes of CO and soot, such as over-firing, closed primary air intake, flame impingement, and lack of combustion air.

✓ Remove dirt, rust, and other debris that may be interfering with the burners. Clean the heat exchanger, if necessary.

✓ Take action to improve draft, if inadequate because of improper venting, obstructed chimney, leaky chimney, or depressurization.

✓ Seal leaks in vent connectors and chimneys.

✓ Adjust gas input if combustion testing indicates overfiring or underfiring.

Measuring BTUH input on Natural Gas Appliances

Use the following procedure when it's necessary to measure and adjust the input of a natural gas appliance.

1. Turn off all gas combustion appliances such as water heaters, dryers, cook stoves, and space heaters that are connected to the meter you are timing, except for the appliance you wish to test.

2. Fire the unit being tested, and watch the dials of the gas meter.

3. Carefully count how long it takes for one revolution of $^1/_2$, 1, or 2 cubic-foot dial. Find that number of seconds in *Table 2-7* in the columns marked "Seconds per Revolution." Follow that row across to the right to the correct column for the $^1/_2$, 1, or 2 cubic-foot dial. Note that you must multiply the number in the table by 1000. Record the input in thousands of BTUs per hour.

4. If the measured input is higher or lower than input on the name plate by more than 10%, adjust gas pressure up or down within a range of 3.2 to 4.0 IWC until the approximately correct input is achieved.

5. If the measured input is still out of range after adjusting gas pressure to these limits, replace the existing orifices with larger or smaller orifices sized to allow the correct input of natural gas.

Table 2-7:   Input in Thousands of BTU/hr for 1000 BTU/cu. ft. Gas

Seconds per Revolution	Size of Meter Dial			Seconds per Revolution	Size of Meter Dial			Seconds per Revolution	Size of Meter Dial		
	1/2 cu. ft.	1 cu. ft.	2 cu. ft.		1/2 cu. ft.	1 cu. ft.	2 cu. ft.		1/2 cu. ft.	1 cu. ft.	2 cu. ft.
15	120	240	480	40	45	90	180	70	26	51	103
16	112	225	450	41	44	88	176	72	25	50	100
17	106	212	424	42	43	86	172	74	24	48	97
18	100	200	400	43	42	84	167	76	24	47	95
19	95	189	379	44	41	82	164	78	23	46	92
20	90	180	360	45	40	80	160	80	22	45	90
21	86	171	343	46	39	78	157	82	22	44	88
22	82	164	327	47	38	77	153	84	21	43	86
23	78	157	313	48	37	75	150	86	21	42	84
24	75	150	300	49	37	73	147	88	20	41	82
25	72	144	288	50	36	72	144	90	20	40	80
26	69	138	277	51	35	71	141	94	19	38	76
27	67	133	267	52	35	69	138	98	18	37	74
28	64	129	257	53	34	68	136	100	18	36	72
29	62	124	248	54	33	67	133	104	17	35	69
30	60	120	240	55	33	65	131	108	17	33	67
31	58	116	232	56	32	64	129	112	16	32	64
32	56	113	225	57	32	63	126	116	15	31	62
33	55	109	218	58	31	62	124	120	15	30	60
34	53	106	212	59	30	61	122	130	14	28	55
35	51	103	206	60	30	60	120	140	13	26	51
36	50	100	200	62	29	58	116	150	12	24	48
37	49	97	195	64	29	56	112	160	11	22	45
38	47	95	189	66	29	54	109	170	11	21	42
39	46	92	185	68	28	53	106	180	10	20	40

Natural Gas Heat Content

Table 2-7 on page 46 assumes that gas is 1000 BTU per cubic foot. Where BTU values differ from this figure—especially at high elevations—obtain the correct BTU value from the gas supplier and apply the formula shown below.

**(BTU value from supplier ÷ 1000) X BTU/hr
input from table = Actual BTU/hr input of**

Gas meter dial: Use the number of seconds per revolution of the one-foot dial and the table on the following page to find the appliance's input.

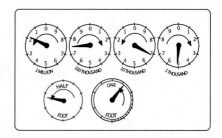

Adjusting Gas Pressure

Use these steps to change gas pressure for the purpose of adjusting gas input to manufacturer's specifications, or to reduce CO production, or to adjust the flue-gas oxygen and steady-state efficiency within acceptable limits.

1. Disable the furnace from firing by turning the thermostat down or turning the main switch off.

2. Remove the gas plug on the burner side of the gas valve and connect a gas-pressure manometer.

3. Remove the cap on the gas-pressure regulator, which is on top of the gas valve.

4. Fire the furnace by turning up the thermostat and activating the main switch.

5. Turn the screw under the cap you removed clockwise to increase pressure and counterclockwise to decrease the gas pressure.

6. Recheck input, CO level, O_2 level and re-adjust if necessary. With a modern flue-gas analyzer you can watch these measurements change as the gas pressure changes.

7. Shut the furnace off, disconnect the manometer, and replace the gas plug and gas-regulator cap.

Measuring gas pressure: Either bellows-type or liquid manometers are used to measure gas pressure.

turning regulator screw

regulator screw cap

Gas Valve Top View

pressure plug

2.4 Evaluating and Improving Oil-Burner Efficiency

Oil burners require annual maintenance to retain their operational safety and combustion efficiency. Testing for combustion efficiency (steady-state efficiency), oxygen, draft, carbon monoxide, and smoke should be used to guide and evaluate maintenance. These procedures apply to oil-fired furnaces, and water heaters.

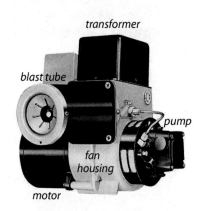

2.4.1 Oil-Burner Inspection

Use visual inspection and combustion testing to evaluate oil burner operation. An oil burner passing visual inspection and giving good test results may need no maintenance. If the test results are fair, adjustments may be necessary. Unsatisfactory test results may indicate the need to replace the burner or the entire heating unit.

Follow these steps to achieve a minimum standard for oil-burner safety and efficiency.

✓ Each oil furnace should have a dedicated electrical circuit. Assure that all 120-volt wiring connections are enclosed in covered electrical boxes.

✓ Verify that all oil-fired heaters are equipped with a barometric draft control, unless they have high-static burners or are mobile home furnaces.

✓ Assure that barometric draft controls are mounted plumb and level and that the damper swings freely.

Barometric draft control: This control supplies a stable over-fire draft and controlled flow of combustion gases through the heat exchanger.

✓ Inspect burner and appliance for signs of soot, overheating, fire hazards, corrosion, or wiring problems.

✓ Inspect fuel lines and storage tanks for leaks.

✓ Inspect heat exchanger and combustion chamber for cracks, corrosion, or soot buildup.

✓ Check to see if flame ignition is instantaneous or delayed. Flame ignition should be instantaneous, except for pre-purge units where the blower runs for a while before ignition.

2.4.2 Oil-Burner Testing

As with gas burners, combustion testing is the key to understanding the oil-fired appliance's current performance level and potential for improvement. Consider the following tests and adjustments for a thorough oil-burner analysis.

✓ Analyze the flue gas for O_2, temperature, CO, and steady-state efficiency (SSE). Sample undiluted flue gases between the appliance and barometric draft control.

✓ Sample undiluted flue gases with a smoke tester, following the smoke-tester instructions. Compare the smoke smudge left by the gases on the filter paper with the manufacturer's smoke-spot scale to determine smoke number.

Table 2-8: Combustion Standards – Oil-Burning Appliances

Oil Combustion Performance Indicator	Non-Flame Retention	Flame Retention
Oxygen (% O_2)	4–9%	4–7%
Stack temperature (°F)	350°–600°	325°–500°
Carbon monoxide (CO) parts per million (ppm)	≤ 100 ppm	≤ 100 ppm
Steady-state efficiency (SSE) (%)	≥ 75%	≥ 80%
Smoke number (1–9)	≤ 2	≤ 1
Excess air (%)	≤ 100%	≤ 25%
Oil pressure pounds per square inch (psi)	≥ 100 psi	≥ 100-150 psi (pmi)*
Over-fire draft (IWC negative)	5 Pa. or .02 IWC	5 Pa. or .02 IWC
Flue draft (IWC negative)	10–25 Pa. or 0.04–0.1 IWC	10–25 Pa. or 0.04–0.1 IWC

* pmi = per manufacturer's specifications

✓Measure over-fire draft over the fire inside the firebox.

✓Measure flue draft between the appliance and barometric draft control.

✓Measure high-limit shut-off temperature and adjust or replace the high-limit control if the shut-off temperature is more than 250° F.

✓Measure oil-pump pressure, and adjust to manufacturer's specifications if necessary.

✓Measure transformer voltage, and adjust to manufacturer's specifications if necessary.

✓Time the CAD cell control or stack control to verify that the burner will shut off, within 45 seconds, when the cad cell is blocked from seeing the flame.

Table 2-9: Minimum Worst-Case Draft for Oil-Fired Appliances

Appliance	Outdoor Temperature (Degrees F)				
	<20	21-40	41-60	61-80	>80
Oil-fired furnace or water heater with atmospheric chimney	−15 Pa. −0.06 IWC	−13 Pa. −0.053 IWC	−11 Pa. −0.045 IWC	−9 Pa. −0.038 IWC	−7 Pa. −0.030 IWC

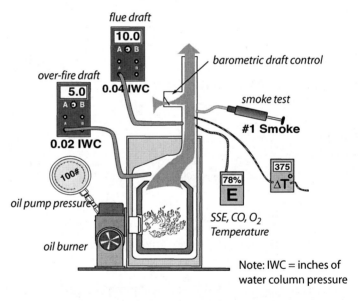

Note: IWC = inches of water column pressure

Measuring oil-burner performance: To measure oil-burning performance indicators, a manometer, flue-gas analyzer, smoke tester, and pressure gauge are required.

2.4.3 Oil-Burner Adjustment

You can adjust fuel, combustion air, and draft on most oil burning appliances. Consider the following maintenance procedures and adjustments.

- ✓ Replace burner nozzle after matching the nozzle size to the home's heat-load requirements.

- ✓ Set oil pump to correct pressure.

- ✓ Adjust air shutter to achieve oxygen and smoke values, specified by the manufacturer or in *Table 2-8 on page 51*.

- ✓ Adjust barometric damper for flue draft of 5–10 pascals or 0.02-to-0.04 IWC (upstream of barometric damper).

- ✓ Adjust fan speed or increase ducted airflow to reduce high flue-gas temperature if appropriate.

2.4.4 Oil-Burner Maintenance and Visual Checks

After evaluating the oil burner's initial operation, perform some or all of the following maintenance tasks as needed to optimize safety and efficiency.

- ✓ Clean the burner's blower wheel.

- ✓ Clean dust, dirt, and grease from the burner assembly.

- ✓ Replace oil filter(s) and clean or replace air filter.

- ✓ Remove soot and sludge from combustion chamber.

- ✓ Remove soot from heat exchange surfaces.

- ✓ Adjust gap between electrodes to manufacturer's specifications.

- ✓ Repair the ceramic combustion chamber, or replace it if necessary.

- ✓ Verify correct flame-sensor operation.

After these maintenance procedures, the technician performs the diagnostic tests described previously to evaluate improvement made by the maintenance procedures and to determine if fine-tuning is required.

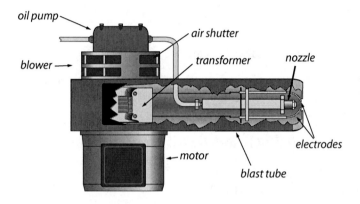

Oil burner: Performance and efficiency will deteriorate over time if neglected. Annual maintenance is recommended.

2.4.5 Upgrading to Flame-Retention Burners

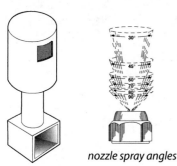

nozzle spray angles

heat exchanger

A flame-retention burner is a newer type of oil burner that gives a higher combustion efficiency by swirling the mist or oil and air to produce better mixing. Flame-retention burners, which have been available for more than 20 years, waste less heat and have steady-state efficiency (SSE) of 80% or slightly more. Replacing an old-style burner with a flame-retention model may be cost-effective if the existing SSE is less than 75%. Flame-retention-burner motors run at

Oil spray pattern and combustion chamber: Matching the burner's spray pattern to the combustion chamber is important to retrofit applications.

3450 rpm and older oil burners run at 1725 rpm motor speed. Looking for the nameplate motor speed can help you discriminate between the flame-retention burners and older models.

If a furnace has a sound heat exchanger but the oil burner is inefficient or unserviceable, the burner may be replaced by a newer flame-retention burner. The new burner must be tested for efficient and safe operation as described previously.

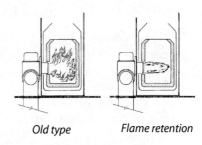

Old type   Flame retention

Flame-retention burner: The burner uses greater air velocity to swirl the oil-air mixture and produces a tighter, hotter flame.

✓ Size the burner and nozzle to match the building's heat load, making adjustments for new insulation and air sealing done during weatherization. (With steam heating, size the burner to existing radiation surface area.)

✓Install new combustion chamber, choosing one that fits the size and shape of the burner flame. Or, change nozzles on the new burner to produce a flame that fits an existing combustion chamber that is still in good condition. Either way, the flame must fill the combustion chamber without impinging to the point where soot is formed.

2.5 Gas Range and Oven Safety

Gas ranges and ovens can produce significant quantities of CO in a kitchen. Overfiring, dirt buildup, and foil installed around burners are frequent causes of CO. Oven burners are likely to produce CO even when not obstructed by dirt or foil. Most range and oven burners are equipped with adjustable needle-and-seat valves. Most ranges also have an adjustable gas regulator that services the entire unit.

2.5.1 Testing Ranges and Ovens

Test the range and oven for safety following these steps and take the recommended actions before or during weatherization.

1. Test each stove-top burner separately, using a digital combustion analyzer or CO meter and holding the probe about 8 inches above the flame.

2. Clean and adjust burners producing more than 25 parts per million (ppm). Burners often have an adjustable gas control.

3. Turn on the oven to bake at high temperature. Sample the CO level in exhaust gases at the oven vent and in the ambient air after 10 minutes. Oven CO level should be less than 100 ppm. The kitchen ambient air CO level should be less than 25 ppm after 10 minutes.

4. Actions to reduce high CO levels include cleaning the oven, removing aluminum foil, or adjusting the burner's adjustable gas control.

5. If the CO reading is over 100 ppm or if the ambient-air reading rises to 25 ppm or more during the test, abort the test.

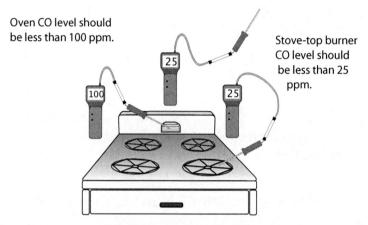

Ambient-air CO level should be less than 25 ppm after 10 minutes.

Oven CO level should be less than 100 ppm.

Stove-top burner CO level should be less than 25 ppm.

2.5.2 CO Reduction Strategies

Gas ranges and ovens are popular and most families won't be replacing them any time soon. You can reduce the risk of CO and other combustion gases by the following.

✓ Verify that the range/oven has a vent fan that exhausts air to outdoors.

✓ Install a CO detector in the kitchen high on a wall away from the range/oven.

✓ Advise customers to open a window slightly when using the range/oven.

✓ Replace the combustion range/oven with an electric model.

2.5.3 Advice to Customers about Range/Ovens

Advise the client of the following important operating practices.

- ✓ Never install aluminum foil around a range burner or oven burner.

- ✓ Never use a range burner or gas oven as a space heater.

- ✓ Open a window and turn on the kitchen exhaust fan when using the range or oven.

- ✓ Keep range burners and ovens clean to prevent dirt from interfering with combustion.

- ✓ Burners should display hard blue flames. Yellow or white flames, wavering flames, or noisy flames should be investigated by a trained gas technician.

- ✓ Buy and install a CO detector, and discontinue use of the range and oven if the CO level rises above 25 ppm in ambient air.

2.6 Inspecting Furnace Heat Exchangers

Leaks in heat exchangers are a common problem, causing the flue gases to mix with house air. Ask clients about respiratory problems, flue-like symptoms, and smells in the house when the heat is on. Also, check around supply registers for signs of soot, especially with oil heating. Furnace heat exchangers should be inspected as part of a service call. Consider using one or more of the following 6 general options for evaluating heat exchangers.

✓ Look for rust at furnace exhaust and vent connector.

✓ Look for flame-damaged areas near the burner flame. Look for flame impingement on the heat exchanger during firing.

✓ Observe flame movement, change in chimney draft, or change in CO reading as blower is turned on and off.

✓ Measure the flue-gas oxygen concentration before the blower starts and just after it has started. There should be no more than a 1% change in the oxygen concentration.

Furnace heat exchangers: Although no heat exchanger is completely airtight, it should not leak enough to display the warning signs described here.

✓ Examine the heat exchanger, shining a bright light on one side and looking for light traces on the other using a mirror or inspection scope to peer into tight locations.

✓ Employ chemical detection techniques, following manufacturer's instructions.

Evaluating Combustion and Venting

CHAPTER 3: VENTING AND COMBUSTION AIR

Venting and combustion air are important to the safe flow of air and gases through the burners, heat exchanger, and chimney of a combustion heating system. Venting is a general term for the flues, chimney and other passageways that exhaust combustion gases out of the home. If the venting system is faulty, combustion gases may spill into the home. If these gases contain carbon monoxide (CO), a poisonous gas, the health of residents is at significant risk.

Depressurization from exhaust fans or a furnace blower often interfere with effective venting. Depressurization can cause combustion gases to backdraft or even cause the flame to roll out of the combustion chamber.

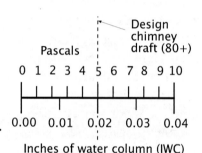

Combustion air provides oxygen necessary for combustion. Most of the time combustion air comes from indoors and is replenished from leaks in the building shell. Sometimes a home or the combustion appliance zone (CAZ) is too airtight to supply adequate combustion air and additional combustion air must be supplied through a dedicated pipe from the outdoors.

3.1 EVALUATING AND IMPROVING VENTING PERFORMANCE

The purpose of evaluating venting performance is to insure that the venting system is venting combustion gases out of a home and that the combustion appliance isn't producing excessive CO. Draft is also an indicator of the effectiveness of the venting system and the stability of the combustion process. Draft is measured in inches of water column (IWC) or pascals. One IWC converts to approximately 250 pascals.

Most existing combustion appliances exhaust their gases into an atmospheric chimney. An atmospheric chimney produces negative draft—a slight vacuum. The strength of this draft is determined by the chimney's height, its cross-sectional area, and the temperature difference between the flue gases and outdoor air. Atmospheric chimney draft pressures should always be negative with reference to the CAZ.

Atmospheric chimneys exhaust combustion gases using the gases' buoyancy. Atmospheric gas appliances are designed to operate at a chimney draft of around negative 0.02 inches of water column (IWC) or –5 pascals. Tall chimneys located indoors typically produce stronger drafts, and short chimneys or outdoor chimneys produce relatively weaker drafts. Wind and house pressures also have a strong influence on draft in atmospheric chimneys.

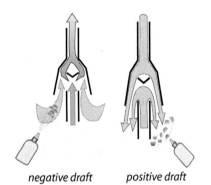

negative draft   *positive draft*

Detecting gas spillage: Spillage results when air flows down the chimney and out the draft diverter.

Fan-assisted appliances employ a small fan near the exhaust of their heat exchanger. This draft-inducing fan regulates the overfire draft but has little or no effect on draft in the atmospheric chimneys.

Positive-draft appliances, like condensing furnaces, have a strong positive draft and an airtight venting system. The positive draft of these appliances is created by a draft fan and is strong enough to resist the influence most indoor and outdoor air pressures.

3.2 VENTING-SYSTEM REQUIREMENTS

Proper venting is essential to the operation, efficiency, safety, and durability of combustion appliances. The National Fire Protection Association (NFPA) and the International Code Council (ICC) are the authoritative information sources on material-choice, sizing, and clearances for venting systems. The information in this venting section is based on the following NFPA and ICC documents.

- The *International Fuel Gas Code* (IFGC) (ICC)
- NFPA 31: *Standard for the Installation of Oil-Burning Equipment*
- NFPA 211: *Standard for Chimneys, Fireplaces, Vents, and Solid-Fuel-Burning Appliances*
- The *International Mechanical Code* (IMC) 2000 edition
- The *International Residential Code* (IRC) 2000 edition

Table 3-1: Guide to Venting Standards

Topic	Standard and Section
Vent Sizing	IFGC, Section 504
Clearances	IFGC, Section 308 and Tables 308.2I NFPA 31, Section 4-4.1.1 and Tables 4-4.1.1 and 4-4.1.2 NFPA 211, Sections 6.5, 4.3, 5
Combustion Air	IFGC, Section 304 IMC, Chapter 7 IRC, Chapter 17 NFPA 31, Section 1-9; NFPA 211, Section 8.5 and 9.3

3.2.1 General Venting Requirements

Combustion gases are vented through vertical chimneys or other types of approved horizontal or vertical vent piping. Identifying the type of existing venting material, verifying the correct size of vent piping, and making sure the venting conforms to the applicable codes are important tasks in inspecting and repairing venting systems. Too large a vent often leads to condensation and corrosion. Too small a vent can result in spillage. The wrong vent materials can corrode or deteriorate from heat.

3.2.2 Vent Connectors

A vent connector connects the appliance's venting outlet with the chimney. Approved vent connectors for gas- and oil-fired units are made from the following materials.

- Type-B vent, consisting of a galvanized-steel outer pipe and aluminum inner pipe

- Type-L vent connector with a stainless-steel inner pipe and either galvanized or black-steel outer pipe.

- Galvanized-steel pipe ($\geq$ 0.019 inch thick or 20 gauge) for vent connectors 5 inches in diameter or less.

- Galvanized-steel pipe ($\geq$ 0.023 inch thick or 22 gauge) for vent connectors 6-to-10 inches in diameter.

Double-wall vent connectors are the best option, especially for appliances with some horizontal vent piping. A double-wall vent connector helps maintain flue-gas temperature and prevent condensation. Gas appliances with draft hoods, installed in attics or crawl spaces must use a Type-B vent connector. Type-L vent pipe is commonly used for vent connectors for oil and solid fuels but can also be used for gas.

Observe the following general specifications for vent connectors.

- A vent connector is almost always the same size as the vent collar on the appliance it vents.

- Single-wall vent-pipe sections should be fastened together with 3 screws or rivets.

- The vent connector should be sealed tightly where it enters a masonry chimney.

- Vent connectors should be free of rust, corrosion and holes.

- The chimney combining two vent connectors should have a cross-sectional area equal to the area of the larger vent connector plus half the area of the smaller vent connector. This common vent should be no larger than 7 times the area of the smallest vent. For specific vent sizes, see NFPA codes themselves listed on *page 63.*

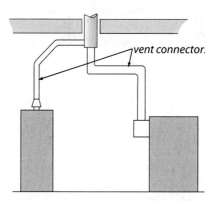

Two vent connectors joining chimney: The water heater's vent connector enters the chimney above the furnace because the water heater has a smaller input.

- The horizontal length of vent connectors shouldn't be more than 75% of the chimney's vertical height or have more than 18 inches horizontal run per inch of vent diameter.

- Vent connectors must have upward slope to their connection with the chimney. A slope of at least $1/4$ inch of rise per foot of horizontal run along their entire length is recommended to cause combustion gases to rise through the vent and to prevent condensation from pooling and rusting the vent.

- When two vent connectors connect to a single chimney, the vent connector servicing the smaller appliance should enter the chimney above the vent for the larger appliance.

- Clearances for common vent connectors are listed in the following table.

Table 3-2: Vent Connector Diam. vs. Max. Horiz. Length

Diameter (in)	3"	4"	5"	6"	7"	8"	9"	10"	12"	14"
Length (ft)	4.5'	6'	7.5'	9'	10.5'	12'	13.5'	15'	18'	21'

From *International Fuel Gas Code 2000*

Table 3-3: Areas of Round Vents

Vent diameter	4"	5"	6"	7"	8"
Vent area (square inches)	12.6	19.6	28.3	38.5	50.2

Table 3-4: Clearances to Combustibles for Vent Connectors

Vent Connector Type	Clearance
Single-wall galvanized-steel vent pipe	6" (gas) 18" (oil)
Type-B double-wall vent pipe (gas)	1" (gas)
Type L double wall vent pipe (stainless steel inner liner, stove pipe or galvanized outer liner)	9", or 1 vent diameter, or as listed

3.2.3 Chimneys

There are two common types of vertical chimneys for venting combustion fuels that satisfy NFPA and ICC codes. First there are masonry chimneys lined with fire-clay tile, and second there are manufactured metal chimneys, including all-fuel metal chimneys and Type-B vent chimneys for gas appliances.

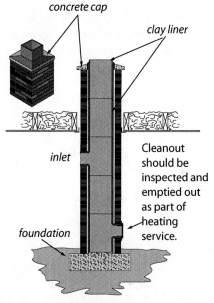

concrete cap

clay liner

inlet

Cleanout should be inspected and emptied out as part of heating service.

foundation

Masonry chimneys: Remain a very common vent for all fuels.

Masonry Chimneys

Observe the following general specifications for building, inspecting, and repairing masonry chimneys.

- Masonry chimneys should be supported by their own masonry foundation.

- Existing masonry chimneys should be lined with a fireclay flue liner. There should be a $^1/_2$-inch to 1-inch air gap between the clay liner and the chimney's masonry to insulate the liner. The liner shouldn't be bonded structurally to the outer masonry because it needs to expand and contract independently of the chimney's masonry structure. The clay liner can be sealed to the chimney cap with a flexible high-temperature sealant.

- The chimney's penetrations through floors and ceilings should be sealed with metal and high-temperature sealant as a firestop and air barrier.

- Deteriorated or unlined masonry chimneys should be rebuilt as specified above or relined as part of a heating-system replacement or a venting-safety upgrade. As an alternative, the vertical chimney may be replaced by a sidewall vent, equipped with a power venter mounted on the exterior wall.

- Masonry chimneys should have a cleanout 12 inches or more below the lowest inlet. Mortar and brick dust should be cleaned out of the bottom of the chimney through the clean-out door, so that this debris won't eventually interfere with venting.

Table 3-5: Clearances to Combustibles for Common Chimneys

Chimney Type	Clearance
Interior chimney masonry w/ fireclay liner	2"
Exterior masonry chimney w/ fireclay liner	1"
All-fuel metal vent: insulated double wall or triple-wall pipe	2"
Type B double-wall vent (gas only)	1"
Manufactured chimneys and vents list their clearance	

Manufactured Chimneys

Manufactured metal chimneys have engineered parts that fit together in a prescribed way. Metal chimneys contain manufactured components from the vent connector to the termination fitting on the roof. Parts include: metal pipe, weight-supporting hardware, insulation shields, roof jacks, and chimney caps. One manufacturer's chimney may not be compatible with another's connecting fittings.

All-fuel metal chimneys come in two types: insulated double wall metal pipe and triple-wall metal pipe. Install them strictly observing the manufacturer's specifications.

All-fuel metal chimney: These chimney systems include transition fittings, support brackets, roof jacks, and chimney caps. The pipe is double- or triple-wall insulated.

Type-B vent pipe is permitted as a chimney for gas appliances. Some older manufactured gas chimneys were made of metal-reinforced asbestos cement.

Chimney Terminations

Masonry chimneys and all-fuel metal chimneys should terminate at least three feet above the roof penetration and two feet above any obstacle within ten feet of the chimney outlet. Chimneys should have a cap to prevent rain and strong downdrafts from entering.

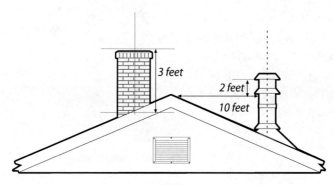

Chimney terminations: Should have vent caps and be given adequate clearance height from nearby building parts. These requirements are for masonry chimneys and manufactured all-fuel chimneys.

B-vent chimneys can terminate as close as one foot above flat roofs and pitched roofs up to a $^6/_{12}$ roof pitch. As the pitch rises, the minimum termination height rises as shown in the table.

Table 3-6: Roof Slope and B-Vent Chimney Height (ft)

flat-6/12	6/12-7/12	7/12-8/12	8/12-9/12	9/12-10/12	10/12-11/12	11/12-12/12	12/12-14/12	14/12-16/12	16/12-18/12
1'	1'3"	1'6"	2'	2'6"	3'3"	4'	5'	6'	7'

From *International Fuel Gas Code 2000*

Metal Liners for Masonry Chimneys

Unlined masonry chimneys or chimneys with deteriorated liners should be relined as part of heating system replacement. Use either Type-B vent, a flexible or rigid stainless-steel liner, or a flexible aluminum liner. *See page 75.*

Flexible liners require careful installation to avoid a low spot at the bottom, where the liner turns a right angle to pass through the wall of the chimney. Follow the manufacturer's instructions, which usually prescribe stretching the liner and fastening it securely at both ends, to prevent it from sagging and thereby creating such a low spot.

Flexible liners are easily damaged by falling masonry debris inside a deteriorating chimney. Use B-vent instead of a flexible liner when the chimney is significantly deteriorated.

Flexible metal chimney liners: The most important installation issues are sizing the liner correctly along with fastening and supporting the ends to prevent sagging.

To minimize condensation, flexible liners should be insulated—especially when installed in exterior chimneys. Consider insulating flexible metal chimney liners with vermiculite or a fiberglass-insulation jackets, if the manufacturer's instructions allow.

Sizing flexible chimney liners correctly is very important. Oversizing is common and can lead to condensation and corrosion. The manufacturers of the liners include vent-sizing tables in their instructions. Liners should bear the label of a testing lab like Underwriters Laboratories (UL).

3.2.4 Special Venting Considerations for Gas

The American Gas Association (AGA) has devised a classification system for venting systems serving natural gas and propane appliances. This classification system assigns Roman numerals to four categories of venting based on whether there is positive or negative pressure in the vent and whether condensation is likely to occur in the vent.

	Negative-pressure Venting	Positive-pressure
Non-condensing	**I** Combustion Efficiency 83% or less Use standard venting: masonry or Type B vent	**III** Combustion Efficiency 83% or less Use only pressurizable vent as specified by manufacturer
Condensing	**II** Combustion Efficiency over 83% Use only special condensing-service vent as specified by manufacturer	**IV** Combustion Efficiency over 83% Use only pressurizable condensing-service vent as specified by manufacturer

American Gas Association Vent Categories

AGA venting categories: The AGA classifies venting by whether there is positive or negative pressure in the vent and whether condensation is likely.

A great majority of appliances found in homes and multifamily buildings are Category I, which have negative pressure in vertical chimneys with no condensation expected in the vent connector or chimney. Condensing furnaces are usually Category IV with positive pressure in their vent and condensation occurring in both the appliance and vent. Category III vents are rare but some fan-assisted appliances are vented with airtight non-condensing vents.

Venting Fan-Assisted Furnaces

Newer gas-fired fan-assisted central heaters control flue-gas flow and excess air better than atmospheric heaters, resulting in their higher efficiency. These are non-condensing Category I furnaces in the 80%-plus Annual Fuel Utilization Efficiency

(AFUE) range. Because these units eliminate dilution air and have slightly cooler flue gases, existing chimneys should be carefully inspected to ensure that they are ready for a possibly more corrosive flue-gas flow. The chimney should be relined when any of the following three conditions are present.

- When the existing masonry chimney is unlined.
- When the old clay or metal chimney liner is deteriorated.
- When the new heater has a smaller input than the old one. In this case the liner should be sized to the new furnace and the existing water heater.

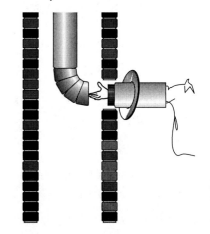

For gas-fired 80+ AFUE furnaces, a chimney liner should consist of:

- Type-B vent
- A rigid or flexible stainless steel liner
- A poured masonry liner
- An insulated flexible aluminum liner

B-vent chimney liner: Double-wall Type-B vent is the most commonly available chimney liner and is recommended over flexible liners. Rigid stainless-steel single-wall liners are also a permanent solution to deteriorated chimneys.

Because of the considerable expense that chimney relining can entail, sidewall venting with a power venter should be considered when an existing chimney is inadequate for new appliances.

Table 3-7: Characteristics of Gas Furnaces

Steady-state efficiency	Operating characteristics
70+	Category I, draft diverter, no draft fan, standing pilot, non-condensing, indoor combustion and dilution air.
80+	Category I, no draft diverter, draft fan, electronic ignition, indoor combustion air, no dilution air.
90+	Category IV, no draft diverter, draft fan, low-temperature plastic venting, positive draft, electronic ignition, condensing heat exchanger, outdoor combustion air is strongly recommended.

Pressurized Sidewall Vents

Sometimes, the manufacturer gives the installer a venting choice of whether to install a fan-assisted furnace into a vertical chimney (Category I) or as a positive-draft appliance (Category III), vented through a sidewall vent. Sidewall-vented fan-assisted furnaces may vent through B-vent, stainless-steel single-wall vent pipe, or high-temperature plastic pipe. Pressurized sidewall vents should be virtually airtight at the operating draft. B-vent must be sealed with high-temperature silicone caulking or other approved means to air-seal its joints.

Some high-temperature positive-draft plastic vent pipe, used in horizontal installations, was recalled by manufacturers because of deterioration from heat and condensation. Deteriorated high-temperature plastic vent should be replaced by airtight stainless-steel vent piping or else B-vent, sealed at joints with high-temperature sealant.

Existing fan-assisted appliances may have problems with weak draft and condensation when vented horizontally. Horizontally vented, fan-assisted furnaces may require a retrofit power venter to create adequate draft in some cases.

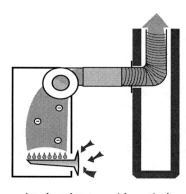 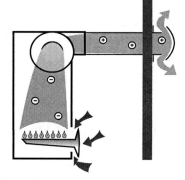

Fan-assisted gas heaters with vertical chimneys: These 80% AFUE central heaters are almost always vented into atmospheric chimneys, which may need to be relined.

Fan-assisted heaters with sidewall vents: Sometimes these appliances are vented through a side wall through airtight plastic or stainless-steel vent pipe.

Condensing-Furnace Venting

Condensing furnaces with 90+ AFUE are vented horizontally or vertically through PVC Schedule 40 pipe. The vent is pressurized and plenty of condensation occurs, making it Category IV. Vent piping should be sloped back toward the appliance, so the condensate can be drained and treated if necessary.

Combustion air is supplied from outdoors through a sealed plastic pipe or from indoors. Outdoor combustion air is highly recommended, and most condensing furnaces are equipped for outdoor combustion air through a dedicated pipe. This combined combustion-air and venting system is referred to as direct-vent or sealed-combustion.

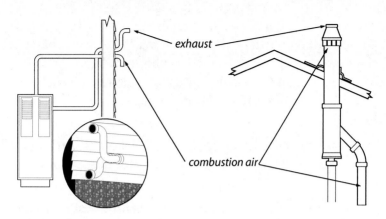

exhaust

combustion air

Condensing furnace venting: The two common types of termination for plastic condensing vents are separate pipes and a combined fitting. Vents going through the roof are preferred for their being more resistant to tampering and damage.

3.2.5 Power Venters for Sidewall Venting

Power venters are installed just inside or outside an exterior wall and are used for sidewall venting. Power venters create a stable negative draft.

Many power venters allow precise control of draft through air controls on the their fans. Barometric draft controls can also provide good draft control when installed either on the common vent for two-appliances or on the vent connector for each appliance. This more precise draft control, provided by the power venter and/or barometric damper, can minimize excess combustion and dilution air. Flue gas temperatures for power ven-

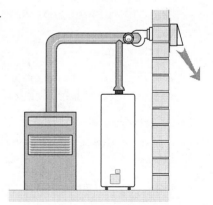

Power venters: Sidewall venting with a power venter is an excellent option when the chimney is dilapidated or when no chimney exists.

ters can be cooler than temperatures needed to move air up vertical atmospheric chimneys. Less excess air and cooler flue gases can improve combustion efficiency in many cases, compared to the non-adjustable draft of a vertical chimney. However, the power venter must be installed by a technician familiar with adjusting the draft to each appliance to achieve the efficiency benefit.

A single power venter can vent both a furnace and also a water heater. Types B or L vent are good choices for horizontal vent piping. Use Type B only for gas.

Power venters should be considered as a venting option when:

- Wind, internal house pressures, or nearby buildings have created a stubborn drafting problem that other options can't solve.

- An existing horizontally vented appliance has weak draft and/or condensation problems.

- Clients who currently heat with electricity want to convert to gas space heating and water heating but have no chimney.

- The cost of lining an unlined or deteriorated chimney exceeds the cost of installing a power venter with its horizontal vent.

- A floor furnace or other appliance with a long horizontal vent connector has backdrafting problems.

- A water heater is orphaned in a too-large vertical chimney when the new furnace is vented through a plastic venting system.

- High draft in the existing vertical chimney is creating unstable combustion or low steady-state efficiency in the appliances connected to it.

3.3 COMBUSTION AIR

Combustion air enters the CAZ through unintentional or intentional openings in the building shell or through a dedicated pipe from outdoors. The need to install an outdoor combustion air source is a common decision in residential HVAC installation and service work. Use worst-case draft testing to help determine whether or not to improve combustion air. *See "Worst-Case Testing for Atmospheric Venting Systems" on page 28.*

The goals of assessing combustion air through draft testing are the following.

- To discover whether or not there is an adequate supply of combustion air.

- To ensure that a combustion-air problem isn't creating CO, weakening draft, or interfering with combustion.

- To avoid unnecessary work or creating a problem by installing combustion air openings. Combustion air openings can depressurize the CAZ in some cases, especially when the openings are above the furnace.

A combustion-air source must deliver between 17 cfm and 600 cfm, depending on the size of the combustion appliances.

Table 3-8: CFM Requirements for Combustion Furnaces

Appliance	Combustion Air (cfm)	Dilution Air (cfm)
Conventional Oil	38	195
Flame-Retention Oil	25	195
High-Efficiency Oil	22	–
Conventional Atmospheric Gas	30	143
Fan-Assisted Gas	26	–
Condensing Gas	17	–
Fireplace (no doors)	100–600	–
Airtight Wood Stove	10–50	–

A.C.S. Hayden, Residential Combustion Appliances: Venting and Indoor Air Quality
Solid Fuels Encyclopedia

A CAZ is classified as either an un-confined space or confined space. An un-confined space is a CAZ connected to enough building air leakage to provide combustion air. A confined space is a CAZ with sheeted walls and ceiling and a closed door that form an air barrier between the appliance and other indoor spaces. For confined spaces, the IFGC prescribes additional combustion air from outside the CAZ. Combustion air is supplied to the combustion appliance in four ways.

1. To an un-confined space through leaks in the building.

2. To a confined space through an intentional opening or openings between the CAZ and other indoor areas where air leaks replenish combustion air.

3. To a confined space through an intentional opening or openings between the CAZ and outdoors or ventilated intermediate zones like attics and crawl spaces.

4. Directly from the outdoors to the combustion appliance through a duct. Appliances with direct combustion-air

ducts are called sealed-combustion or direct-vent appliances.

3.3.1 Un-Confined-Space Combustion Air

Combustion appliances located in most basements, attics, and crawl spaces get adequate combustion air from leaks in the building shell. Even when a combustion appliance is located within the home's living space, it usually gets adequate combustion air from air leaks unless the house is airtight or the combustion zone is depressurized.

3.3.2 Confined-Space Combustion Air

A confined space is defined by the IFGC as a room containing one or more combustion appliances that has less than 50 cubic feet of volume for every 1000 BTU per hour of appliance input.

However, if a small mechanical room is connected to adjacent spaces through large air passages like floor-joist spaces, the CAZ may not need additional combustion air despite sheeted walls and a door separating it from other indoor spaces. The extent of the connection between the CAZ and other spaces can be confirmed by worst-case draft testing or blower-door pressure testing.

On the other hand, if the home is unusually airtight, the CAZ may be unable to provide adequate combustion air, even when the combustion zone is larger than the minimum confined-space room volume, defined above.

Combustion air from adjacent indoor spaces is usually preferred over outdoor combustion air because of the possibility of wind depressurizing the combustion zone. However, if there is a sheltered outdoor space from which to draw combustion air, outdoor combustion air may be a superior choice. Outdoor air is generally cleaner and dryer than indoor air, and a connection to the outdoors makes the confined space less affected by indoor pressure fluctuations.

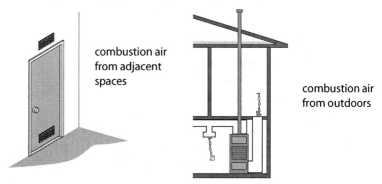

combustion air from adjacent spaces

combustion air from outdoors

Passive combustion-air options: Combustion air can be supplied from adjacent indoor spaces or from outdoors. Beware of passive combustion-air vents into the attic that could depressurize the combustion zone or allow moisture to travel into the attic.

In confined spaces or airtight homes where outdoor combustion air is needed, prefer a single vent openings installed as low in the CAZ as practical. A combustion-air vent into an attic may depressurize the combustion zone or dump warm moist air into the attic. Instead, connect the combustion zone to a ventilated crawl space or directly to outdoors through a single low vent if possible.

Choose an outdoor location that is sheltered, where the wall containing the vent isn't parallel to prevailing winds. Wind blowing parallel to an exterior wall or at a right angle to the vent opening tends to de-pressurize both the opening and the CAZ connected to it. Indoors, locate combustion air vents away from water pipes to prevent freezing in cold climates.

Net free area is smaller than actual vent area and takes the blocking effect of louvers into account. Metal grills and louvers provide 60% to 75% of their area as net free area while wood louvers provide only 20% to 25%. Combustion air vents should be no less than 3 inches in their smallest dimension.

Table 3-9: Combustion Air Openings: Location and Size

Location	Dimensions
Two direct openings to adjacent indoor space	Minimum area each: 100 in^2 1 in^2 per 1000 BTUH each Combined room volumes must be $\geq$ 50 ft^3/1000 BTUH
Two direct openings or vertical ducts to outdoors	Each vent should have 1 in^2 for each 4000 BTUH
Two horizontal ducts to outdoors	Each vent should have 1 in^2 for each 2000 BTUH
Single direct or ducted vent to outdoors	Single vent should have 1 in^2 for each 3000 BTUH

From the International Fuel Gas Code

Here is an example of sizing combustion air to another indoor area. The furnace and water heater are located in a confined space. The furnace has an input rating of 100,000 BTU/hour. The water heater has an input rating of 40,000 BTU/hour. Therefore, there should be 280 in^2 of net free area of vent between the mechanical room and other rooms in the home. ([100,000 + 40,000] $\div$ 1,000 = 140 x 2 in^2 = 280 in^2). Each vent should therefore have a minimum of 140 in^2.

Direct Outdoor Combustion-Air Supply

Many new combustion appliances are designed for direct outdoor-air supply to the burner. These include most condensing furnaces, mobile home furnaces, mobile home water heaters, many space heaters, and some non-condensing furnaces. Some appliances give installers a choice between indoor and outdoor combustion air. Outdoor combustion air is usually preferable in order to prevent the depressurization problems, combustion-air deficiencies, and draft problems.

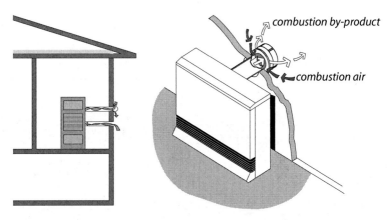

combustion by-product

combustion air

Sealed combustion: Sealed combustion appliances draw combustion air in and exhaust combustion by-products, either using a draft fan or by pressure differences created by the fire.

Fan-Powered Combustion Air

At least one company manufactures a proprietary combustion-air system that introduces outdoor air through a fan that sits on the floor and attaches to a combustion-air duct to outdoors.

Combustion Air Combined with Power Venting

Both gas- and oil-fired heating systems can be supplied with combustion air by proprietary systems that combine power

combustion-air fans

Fan-powered combustion air: Fans for supplying combustion air can help solve stubborn combustion air and drafting problems.

venting with powered combustion-air supply. The combustion air simply flows into the combustion zone from outdoors, powered by the power venter. If the appliance has a power burner, like a gun-type oil burner, a boot may be available to supply combustion air directly to the burner as shown here.

CHAPTER 4: HEATING SYSTEM INSTALLATION

Replacement furnaces should have a minimum Annual Fuel Utilization Efficiency (AFUE) of 80%. However prefer gas furnaces with AFUEs of 90% or more. When high-efficiency furnaces are installed as sealed-combustion units, they provide health and safety benefits in addition to their superior efficiency.

Don't assume that older furnaces are inefficient without testing them. During testing, make appropriate efforts to repair and adjust the existing furnace before deciding to replace it. Replacement parts for older heating units are commonly available.

Furnaces are often replaced when the cost of repairs and retrofits exceeds one half of estimated replacement costs. Estimate the repair and retrofit costs and compare them to replacement cost before deciding between retrofit and replacement.

New heating appliances must be installed to manufacturer's specifications, following all applicable electrical, plumbing, mechanical, and fire codes.

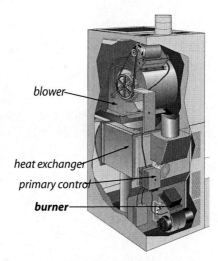

Oil-fired downflow furnaces: Their design hasn't changed much in recent years except for the flame-retention burner.

Heat load calculations, used to size the new heater, should account for reduced heating loads, resulting from insulation and air-sealing work. Heat load calculations should follow Manual J procedures.

4.1 GAS-FIRED HEATING INSTALLATION

The goal of appliance replacement is to save energy and provide safer heating. The heating replacement project should produce a gas-fired heating system in virtually new condition, even though existing components like the gas lines, chimney, water piping, or ducts may remain. Any necessary maintenance or repair on these remaining components must be part of the installation. Any design flaws in the original system should be diagnosed and corrected during the heating-system replacement.

- ✓A new furnace should have an Annual Fuel Utilization Efficiency (AFUE) of at least 90% and be sealed-combustion, have electronic ignition, and have a condensing heat exchanger.

- ✓New oil-fired furnaces should be installed with adequate clearances to facilitate maintenance.

- ✓Clock gas meter to measure gas input and adjust gas input if necessary. *See page 45.*

- ✓Follow manufacturer's venting instructions along with the NFPA 54 to establish a proper venting system. *See "General Venting Requirements" on page 63.*

- ✓Check clearances of heating unit and its vent connector to nearby combustibles, according to the National Fuel Gas Code (NFPA 54). *See page 63.*

Testing New Gas-Fired Heating Systems

- ✓Perform combustion testing and adjust fuel-air mixture to minimize O_2, without creating CO.

- ✓Test the new venting system for CO. *See pages 26 and 34.*

- ✓Test gas water heater to insure that it vents properly after installation of a sealed-combustion, 90+ AFUE furnace. Install a chimney liner if necessary.

Table 4-1: Gas-Burning Furnace and Boiler Combustion Standards

Gas Combustion Performance Indicator	80+	90+
Oxygen (% O_2)	4–8%	4–8%
Stack temperature (°F)	325°–450°	90°–120°
Carbon monoxide (CO) parts per million (ppm)	≤ 100 ppm	≤ 100 ppm
Steady-state efficiency (SSE) (%)	80–83%	92–97%
Gas pressure (inches water column or IWC)	3.2–3.9 IWC	3.2–3.9 IWC

4.2 OIL-FIRED HEATING INSTALLATION

The goal of the system replacement is to save energy and provide safer, more reliable heating. System replacement should provide an oil-fired heating system in virtually new condition, even though components like the oil tank, chimney, piping, or ducts may remain. Any maintenance or repair on these remaining components should be part of the job. Any design flaws related to the original system should be diagnosed and corrected during the heating-system replacement.

✓New oil-fired furnaces should have an AFUE of 80% or more.

✓New oil-fired furnaces should be installed with adequate clearances to facilitate maintenance.

✓Examine existing chimney and vent connector for suitability as venting for new appliance. The vent connector may need to be re-sized and the chimney may need to be re-lined.

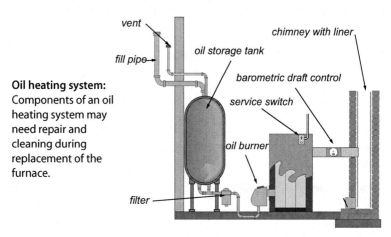

Oil heating system: Components of an oil heating system may need repair and cleaning during replacement of the furnace.

Labels in figure: vent, fill pipe, oil storage tank, chimney with liner, barometric draft control, service switch, oil burner, filter

✓Check clearances of heating unit and its vent connector to nearby combustibles, by referring to NFPA 31. See *"Clearances to Combustibles for Vent Connectors" on page 66.*

✓Install new fuel filter and purge fuel lines as part of new installation.

✓Inspect oil tank and remove deposits at bottom of tank.

✓Bring tank and oil lines into compliance with NFPA 31.

✓Verify the presence of emergency shut-off, installed in the living space.

Testing New Oil-Fired Heating Systems

✓Check for the presence of a control that interrupts power to the burner in the event of a fire.

✓Measure oil pressure and note the nozzle's gallon-per-minute (gpm) rating. Adjust oil pressure or replace nozzle as necessary to achieve correct input. Verify input from oil pressure and nozzle gpm, using nozzle manufacturer's data.

✓Verify correct spray angle and spray pattern.

✓Test transformer voltage to verify compliance with manufacturer's specifications.

✓ Test control circuit amperage, and adjust thermostat heat anticipator to match. Or follow thermostat manufacturer's instructions for adjusting cycle length.

✓ Adjust oxygen, flue-gas temperature, and smoke number to match manufacturer's specifications or specifications given here. Smoke number should be zero on all modern oil-fired equipment.

Table 4-2: Combustion Standards – Oil-Burning Appliances

Oil Combustion Performance Indicator	Standard
Oxygen (% O_2)	4–7%
Stack temperature (°F)	325°–450°
Carbon monoxide (CO) parts per million (ppm)	≤ 100 ppm
Steady-state efficiency (SSE) (%)	≥ 80%
Smoke number (1–9)	≤ 1
Excess air (%)	≤ 25%
Oil pressure pounds per square inch (psi)	≥ 100-150 psi (pmi)*
Over-fire draft (IWC negative)	5 Pa. or .02 IWC
Flue draft (IWC negative)	10–25 Pa. or 0.04–0.1 IWC

* pmi = per manufacturer's instructions

4.3 COMBUSTION FURNACE REPLACEMENT

The goal of furnace replacement is to provide a forced-air heating system in virtually new condition, even though existing supply and return ducts may remain. Any existing flaws in the ducts and registers should be diagnosed and corrected during the furnace replacement.

Observe the following standards in furnace installation.

Sealed combustion heaters: Sealed combustion furnaces prevent the air pollution and house depressurization caused by some open-combustion heating

✓ Furnace should be sized to the approximate heating load of the home, accounting for post-weatherization heat-loss reductions.

✓ Installer should add return ducts or supply ducts as part of furnace replacement to improve air distribution, to eliminate duct-induced house pressures, and to establish acceptable values for static pressure and temperature rise.

✓ Seal holes through the jacket of the air handler with mastic or foil tape.

✓ Supply and return plenums should be mechanically fastened with screws and sealed to air handler to form an airtight connection on all sides of these important joints.

✓ All ducts should be sealed as described on page 139.

✓ Set thermostat's heat anticipator to the amperage measured in the control circuit, or follow thermostat manufacturer's instructions for adjusting cycle length.

✓ Temperature rise (supply temperature minus return temperature) must be within manufacturer's specifications.

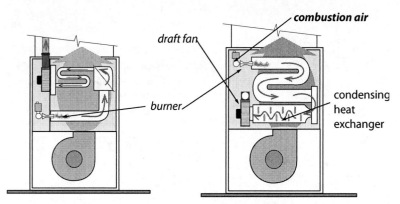

80+ gas furnace: An 80+ furnace has a restrictive heat exchanger, a draft fan (draft inducer), and has no draft diverter or standing pilot.

90+ gas furnace: A 90+ furnace has a condensing heat exchanger and a stronger draft fan for pulling combustion gases through its more restrictive heat exchange system and establishing a strong positive draft.

✓High limit should stop fuel flow at 250° F or less. Furnace must not cycle on high limit.

✓Fan control should be adjusted to activate fan at 130° to 140° F and deactivate it at 95° to 105°F. Higher temperature settings are acceptable if these recommended settings cause a comfort complaint.

✓Blower should not be set to operate continuously.

✓Total external static pressure (TESP) should be within manufacturer's specifications.

✓Filters should be held firmly in place and provide complete coverage of blower intake or return register. Filters should be easy to replace.

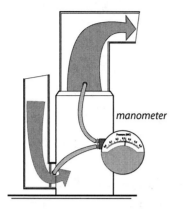

manometer

Static pressure and temperature rise: Testing static pressure and temperature rise across the new furnace should verify that the duct system isn't restricted. The correct airflow, specified by the manufacturer, is necessary for high efficiency.

4.4 Gas Space-Heater Replacement

Space heaters are inherently more efficient than central heaters, because they have no distribution system. As homes become more airtight and better insulated, space heaters become a more practical option for heating the whole home.

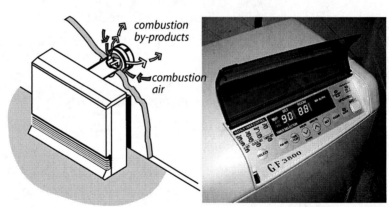

combustion by-products

combustion air

Sealed Combustion Space Heater: These heaters draw combustion air in and exhaust combustion by-products, using a draft fan.

Space Heater Controls: Many modern energy-efficient space heaters have programmable thermostats as standard equipment.

Install space heaters as an energy conservation measure or for health and safety reasons — to replace an unsafe furnace, for example. Use the highest efficiency unit available for the application.

✓ Follow manufacturer's venting instructions carefully. Don't vent sealed-combustion, induced-draft space heaters into naturally drafting chimneys.

✓ Verify that flue-gas oxygen and temperature are within the ranges specified in *Table 2-5 on page 39*.

✓ If the space heater sits on a carpeted floor, specify a fire-rated floor protector, sized to the width and length of the space heater, as a base.

✓ Locate space heater away from traffic, draperies, and furniture.

✓ Provide the space heater with a properly grounded duplex receptacle for its electrical service.

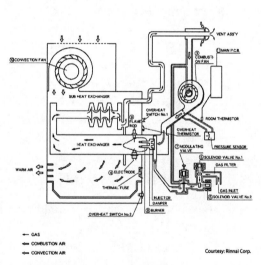

Energy-efficient space heater: Modern space heaters combine sealed combustion, effective heat exchange and forced-air circulation.

Courtesy: Rinnai Corp.

4.4.1 Space-Heater Operation

Inform the customer of the following operating instructions.

✓ Don't store any objects near the space heater that would restrict airflow around it.

✓ Don't use the space heater to dry clothes or for any purpose other than heating the home.

✓ Don't allow anyone to lean or sit on the space heater.

✓ Don't spray aerosols near the space heater. Many aerosols are flammable or can cause corrosion to the space heater's heat exchanger.

4.4.2 Un-vented Space Heaters

Un-vented space heaters are common in some regions of the Southern U.S. These un-vented space heaters deliver all their products of combustion to the indoors. They are not a safe heating option and should be replaced with vented space heaters or electric space heaters.

4.5 ELECTRIC HEATING

Electricity is a cleaner, more convenient form of energy than gas or other fuels, but it is considerably more expensive. Electric heaters are usually 100% efficient at converting the electricity to heat in the room where they are located. However, coal- or oil-generated electricity converts only about 30% of the fuel's potential energy to electricity.

4.5.1 Electric Baseboard Heat

Electric baseboard heaters are zonal heaters controlled by thermostats within the zone they heat. Baseboard heaters contain electric resistance heating elements encased in metal pipes. These pipes extend the length of the unit and are surrounded by aluminum fins to aid heat transfer. As air within the heater is heated, it rises into the room. This draws cooler air into the bottom of the heater.

✓ Make sure that the baseboard heater sits at least an inch above the floor to facilitate good air convection.

✓ Clean fins and remove dust and debris from around and under the baseboard heaters as often as necessary.

✓ Avoid putting furniture directly against the heaters. To heat properly, there must be space for air convection.

There are two kinds of built-in electric baseboard heaters: strip-heat and liquid-filled. Strip-heat units are less expensive than liquid-filled, but they don't heat as well. Strip-heat units release heat in short bursts, as the temperature of the heating elements rises to about 350°F. Liquid-filled baseboard heaters release heat more evenly over longer time periods, as the element temperature rises only to about 180°F.

The line-voltage thermostats used with baseboard heaters sometimes do not provide good comfort. This is because they allow the temperature in the room to vary by 2°F or more. Newer, more accurate thermostats are available. Automatic setback thermostats for electric baseboard heat employ timers or a resident-activated button that raises the temperature for a time and then automatically returns to setback.

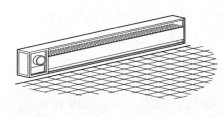

Electric baseboard: Electric baseboard is more efficient than an electric furnace and sometimes even outperforms a central heat pump because it is easily zoneable. The energy bill is determined by the habits of the occupants and the energy efficiency of the building.

4.5.2 Electric Furnaces

Electric furnaces are obsolete heating devices because they are so expensive and environmentally destructive to operate. An electric furnace heats air moved by its fan over several electric-resistance heating elements. Electric furnaces have three to six elements — 3.5 to 7 kW each — that work like the elements in a toaster. The 24-volt thermostat circuit energizes devices called sequencers that bring the 240 volt heating elements on in stages when the thermostat calls for heat. The variable speed fan switches to a higher speed as more elements engage to keep the air temperature stable.

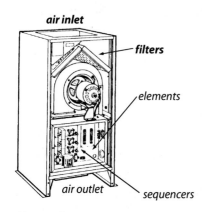

air inlet

filters

elements

air outlet

sequencers

Electric furnace: A squirrel-cage blower blows air over 3 to 6 electric resistance coils and down into the plenum below the floor.

Electric furnaces can be a problem for utility companies if they are using more 5-kW heating elements than are necessary to heat the home—the utility has a higher peak demand than it would if only the minimum number of elements were used. During mild weather, a couple elements are needed and in severe weather they might all be needed.

A standard heating thermostat, combined with an outdoor thermostat, can be used to stage heating elements for different weather. This is not an energy-saving measure but a power-saving measure. Since staging elements benefit the utility company, they may be willing to pay for the savings to the utility power system.

Because electric furnaces are probably the most expensive way to heat a building, observe these specifications.

✓ Replace air filters at regular intervals. The electric heating elements should be dusted and vacuumed if they are dirty. However, cleaning the heating elements shouldn't be necessary if filters are changed regularly.

✓ Seal ducts absolutely airtight and insulate supply ducts.

✓ Install an outdoor staging thermostat to reduce peak load and/or increase comfort.

✓ Replace the electric furnace with another heating source.

4.5.3 Room Heat Pumps

Room heat pumps can provide all or part of the heating and cooling needs for small homes in mild or warm climates. These one-piece room systems look like a room air conditioner, but provide heating as well as cooling. They can also provide ventilation air when neither heating nor cooling are required. They mount in a window or through a framed opening in a wall.

Room heat pumps can be a good choice for replacing existing un-vented gas space heaters or obsolete central heating systems. Their fuel costs may be somewhat higher than oil or gas furnaces, though they are safer and require less maintenance than combustion appliances. Room heat pumps also gain some overall efficiency because they heat a single zone and don't have the delivery losses associated with central furnaces and ductwork. If they replace electric resistance heat, they consume only one-half to one-third the electricity to produce the same amount of heat.

Room heat pumps have a cooling efficiency comparable to the best new window air conditioners. They operate at up to twice the efficiency of older air conditioners.

Room Heat-Pump Installation

Room heat pumps draw a substantial electrical load, and may require 240-volt wiring. Provide a dedicated circuit that can support the equipment's rated electrical input. Insufficient wiring capacity can result in dangerous overheating, tripped circuit breakers, blown fuses, or motor-damaging voltage drops. In most cases a licensed electrician should confirm that the house wiring is sufficient. Don't run portable heat pumps or any other appliance with extension cords or plug adapters.

Room heat pump: Usually installed through an exterior wall, room heat pumps are more efficient than central heat pumps because they have no ducts.

Observe the following specifications when installing room heat pumps.

Table 4-3:   Installing Room Heat Pumps of Air Conditioners

Issue	Window installation	Wall installation
Difficulty of installation	Easiest.	Wall framing required.
Access issues	Window will be inoperable, with no access for fire egress or ventilation.	None.
Air sealing	Care required to properly seal unit to window jambs.	Easy to seal permanently.
Future adaptability	Easy to remove if homeowner switches fuel or type of system.	Poor. Homeowner is left with a hole in the wall if they switch fuel or type of system.

✓Install the unit in a central part of the home where air can circulate to other rooms. Choose a location near an electrical outlet, or where a new outlet can be installed if it's needed.

✓Don't install the unit where bushes will interfere with its outdoor airflow. Heat pumps need lots of outdoor air circulation to operate at maximum efficiency.

✓If you install the unit in a window, choose a double-hung or sliding window that stores out of the way. Portable units don't work well in out-swinging casement windows or up-swinging awning windows.

✓If you install the unit in a framed opening in the wall, use the same guidelines you would to frame a new window or door. Provide headers, beams, or other structural supports where studs are cut, or install it in an opening under an existing window where structural support is already provided by the window framing.

✓ Provide solid supports underneath the unit. These can be manufactured brackets, wood-framed brackets, or brackets fabricated from metal. Fasten the unit with screws to the window jamb and/or sash.

✓ Seal around the exterior siding and trim to keep rain out of the wall cavity. Seal the unit to the opening with the shields provided by manufacturer or with plywood, caulking, or sheet metal.

4.6 WOOD STOVES

Wood heating is a popular and effective auxiliary heating source for homes. However, wood stoves and fireplaces can cause indoor-air-pollution and fire hazards. It's important to inspect wood stoves to evaluate potential hazards.

4.6.1 Wood Stove Clearances

Stoves that are listed by a testing agency like Underwriters Laboratory have installation instructions stating their clearance from combustibles. Unlisted stoves must adhere to clearances specified in NFPA 211.

Stove Clearances

Unlisted stoves must be at least 36 inches away from combustibles. However, listed wood stoves may be installed to as little as 6 inches away from combustibles, if they incorporate heat shields and combustion design that directs heat away from the back and sides. Ventilated or insulated wall protectors may also decrease unlisted clearance from one-third to two thirds. Always follow the stove manufacturer's or heat-shield manufacturer's installation instructions.

Floor Construction and Clearances

Wood stoves must rest of a floor of non-combustible construction. An example of a non-combustible floor is one composed of only masonry material sitting on dirt. This floor must extend no less than 18 inches beyond the stove in all directions. Approved floor protectors or the stove-bottom heat shields of listed stoves can allow the stove to rest on a floor containing combustible materials. The floor would need a minimum of one-quarter inch of grouted tile or an approved floor protector extending 18 inches away from the stove in all directions.

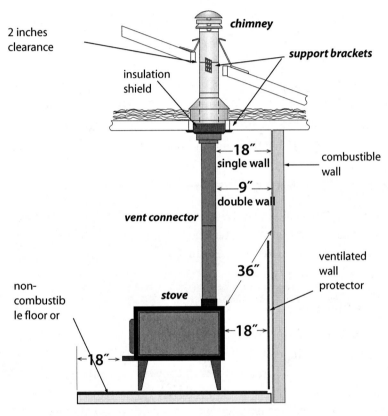

Wood-Stove Installation: Wood-stove venting and clearances are vitally important to wood-burning safety.

Vent-Connector and Chimney Clearance

Interior chimneys require a 2-inch clearance from combustibles and exterior chimneys require a 1-inch clearance from combustibles.

Single-wall vent connectors must be at least 18 inches from combustibles. Wall protectors may reduce this clearance up to two-thirds. Type-L double-wall vent pipe requires only a 9-inch clearance from combustibles.

See also *"Chimneys" on page 67* and *"Vent Connectors" on page 64*.

4.6.2 Wood Stove Inspection

All components of wood-stove venting systems should be approved for use with wood stoves. Chimney sections penetrating floor, ceiling, or roof should have approved thimbles, support packages, and ventilated shields to protect combustible materials from high temperatures. The energy auditor should perform or specify the following inspection tasks, depending on the customer's instructions and the work scope of the energy program.

✓ Inspect stove, vent connector, and chimney for correct clearances from combustible materials as listed in NFPA 211.

✓ If the home is tight, the wood stove should be equipped with outdoor combustion air.

✓ Galvanized steel pipe must not be used to vent wood stoves.

✓ Inspect vent connector and chimney for leaks. Leaks should be sealed with a high-temperature sealant designed for sealing wood-stove vents.

✓ Inspect chimney and vent connector for creosote build-up, and suggest chimney cleaning if creosote build-up exists.

✓ Inspect the house for soot on seldom-cleaned horizontal surfaces. If soot is present or if the blower door indicates leakage, inspect the wood-stove door gasket. Suggest sealing stove air leaks and improving drafting order to reduce indoor smoke emissions.

✓ Inspect stack damper and/or combustion air intake.

✓ Check catalytic combustor for repair or replacement if the wood stove has one.

✓ Assure that heat exchange surfaces and flue passages within the wood stove are free of accumulations of soot or debris.

✓ Wood stoves installed in mobile or manufactured homes must be approved for use in mobile or manufactured homes.

4.7 PROGRAMMABLE THERMOSTATS

A programmable thermostat may be a big energy saver if the building occupant understands how to program it. A programmable thermostat won't save any energy if occupants already control day and night temperatures effectively.

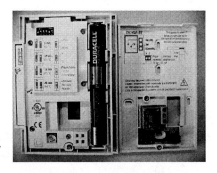

Inside a Programmable Thermostat: In addition to the instructions on the exterior of this thermostat are instructions inside for setting the heat anticipator.

If the existing thermostat is replaced as a part of weatherization or home performance work, discuss programmable thermostats with occupants. If they are willing to use a programmable thermostat, proceed with the installation. Train occupants on the use of the thermostat and leave a copy of manufacturers directions with them.

Many models of programmable thermostats have settings that are selected from inside the thermostat. These include the heat anticipator setting, which adjusts the cycle length of the heating or cooling system.

Chapter 5: Evaluating Forced-Air System Performance

The annual system efficiency of forced-air heating and air-conditioning systems is affected by duct leakage, system airflow, blower operation, balance between supply and return air, and duct insulation levels. The forced-air system usually offers more opportunity for energy savings and comfort improvement than improving combustion or refrigeration equipment.

The testing and evaluation, presented here, has an important sequence. First, deal with the airflow problems because you might have to change the duct system substantially. Then test the ducts for leakage and evaluate whether they are located within the thermal boundary or not. Decide whether duct sealing is important and if so, find and seal the leaks. Finally, if supply ducts are outside the thermal boundary, insulate them. The following list summarizes this logical sequence.

1. Evaluate and/or measure airflow.

2. Troubleshoot specific airflow problems.

3. Make necessary airflow improvements.

4. Evaluate and/or measure duct air leakage.

5. Find duct leaks and seal them.

6. Consider insulating supply ducts and possibly return ducts.

This chapter's first section tells how to evaluate the forced-air system's performance with tables and simple measurements. The second section outlines troubleshooting procedures. The third sections presents procedures for measuring airflow and duct air leakage. The last sections cover air-sealing and insulating ducts.

5.1 Evaluating Forced-Air-System Airflow

Airflow evaluation and improvement comes before duct-sealing because you wouldn't want to make substantial modifications to the ducts you just air-sealed with mastic. The simple airflow evaluation techniques described in this section are less time-consuming than measuring airflow as described later in this chapter.

The most accurate and reliable methods for measuring system airflow are the duct-blower method and the flow-plate method. Measuring return airflow with a flow hood is sometimes accurate if the flow hood is properly calibrated and used according to manufacturer's instructions. These measuring techniques are discussed in *"Measuring Airflow" on page 118*.

5.1.1 Evaluating Duct Design and Recommended Airflow

The air handler's recommended airflow depends on its heating or cooling capacity. For combustion furnaces, there should be 11-to-15 cfm of airflow for each 1000 BTUH of output. To provide this airflow, the supply duct and return duct, connected to the air handler, should have at least 2 square inches of cross-sectional area for each 1000 BTUH of furnace input.

Central air conditioners and heat pumps should deliver 400 cfm ±20% of airflow per ton of cooling capacity (one ton equals 12,000 BTUs per hour). This airflow is usually facilitated by a duct system with at least 6 square inches of cross-sectional area of both supply duct and return duct, connected to the air handler, for each 1000 BTUH of air-conditioning capacity.

In dry climates, you may increase performance and efficiency by increasing airflow to 480 cfm per ton if noise and comfort allow. In humid climates, the recommended airflow per ton may be less than 400 cfm per ton to facilitate dehumidification by keeping the coil cooler

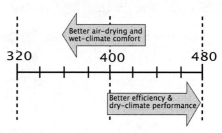

Airflow and climate: More airflow per ton provides better efficiency and performance for dry climates, and less airflow provides better dehumidification for wet climates.

and air moving more slowly across the coil than in a drier climate.

The most accurate way to size new ducts or to evaluate existing ducts is with a duct-sizing computer program, which tells you the size of new ducts or whether existing ducts are adequately sized.

Table 5-1: Recommended Cross-Sectional Area of Metal Supply and Return Ducts at Air Handler

Gas Furnaces		Air Conditioners	
BTUH Input	In^2 Area (Supp. & Ret.)	BTUH Capacity	In^2 Area (Supp. & Ret.)
40,000	80	24,000	144
60,000	120	30,000	180
80,000	160	36,000	216
100,000	200	42,000	252
120,000	240	48,000	288
140,000	280	54,000	324
160,000	320	60,000	360

Each trunk, supply and return, should have the recommended cross-sectional area shown here. Courtesy Delta-T Inc.

Table 5-2:   Round-Duct Square-Inch Equivalency for Metal Ducts

Diameter	Square Inches	Diameter	Square Inches
5	20	12	113
6	28	14	154
7	38	16	201
8	50	18	254
9	64	20	314
10	79	22	380

For flex duct, use the next largest size to get a similar airflow as through round metal duct. Courtesy Delta-T Inc.

Airflow and Blower Speed

A blower can have as many as five speeds. If the air handler's specifications are available, check whether the blower is providing adequate airflow for heating and cooling. Heating typically uses a lower speed and cooling uses a higher one. Each speed represents a different airflow depending on the Total External Static Pressure (TESP) created by the forced-air system.

If the blower speed isn't obvious when looking at the air-handler terminal block, clamp an ammeter around the colored wires until you identify the colors corresponding to the heating and cooling modes.

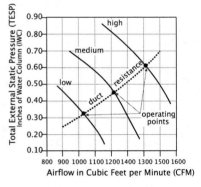

Fan curves: This common type of graph represents the relationship of the blower with the connected ducts. As TESP increases, fan flow decreases as shown by the 3 fan curves. And as airflow increases, TESP increases as shown by the duct-resistance curve. Where the resistance curve meets the fan curves are the operating points, which are your 3 choices of airflow for this air handler.

Solving Obvious Airflow Problems

You probably don't need sophisticated test instruments to discover dirty blowers and coils or disconnected branch ducts. Finding these problems before measuring duct airflow accelerates measurement, troubleshooting, and duct sealing. The following steps precede airflow measurements.

1. Ask the customer about comfort problems and temperature differences in various parts of the home.

2. Based on the customers comments, look for disconnected, restricted ducts, and other obvious problems.

3. Inspect the filter(s), blower, and indoor coil for dirt. Clean them if necessary. If the indoor coil isn't easily visible, a dirty blower is a fair indicator that the coil may also be dirty.

4. Inspect for dirty or damaged supply and return grills that restrict airflow. Clean and repair them.

5. Look for closed registers or balancing dampers that could be restricting airflow to rooms.

6. Notice moisture problems like mold and mildew. Moisture sources, like a wet crawl space, can overpower air conditioners by introducing more moisture into the air than the air conditioner can remove.

5.1.2 Evaluating Furnace Performance

Furnace efficiency depends on temperature rise, fan-control temperatures, and flue-gas temperature. For efficiency, you want a low temperature rise. However, you must maintain a minimum flue-gas temperature to prevent corrosion in the venting of naturally drafting combustion furnaces. Apply the following furnace-operation standards to maximize the heating system's seasonal efficiency and safety.

✓ Perform a combustion analysis as described in *"Combustion Standards for Gas Furnaces" on page 39*. Apply the combustion standards listed in that table to your testing and adjustment.

✓ Check temperature rise after 5 minutes of operation. Refer to manufacturer's nameplate for acceptable temperature rise (supply temperature minus return temperature). The temperature rise should be between 40°F and 70°F with the lower end of this scale being preferable for energy efficiency.

✓ The fan-off temperature should be between 95° and 105° F, with the lower end of the scale being preferable for maximum efficiency.

✓ The fan-on temperature should be 120-140° F, the lower the better.

✓ On time-activated fan controls, verify that the fan is switched on within two minutes of burner ignition and is switched off within 2.5 minutes of the end of the combustion cycle.

✓ The high-limit controller should shut the burner off before the furnace temperature reaches 200°F.

✓ All forced-air heating systems must deliver supply air and collect return air only within the intentionally heated portion of the house. Taking return air from an un-heated area of the house such as an unoccupied basement is not acceptable.

✓ There should be a strong noticeable airflow from all supply registers.

✓ The blower shouldn't operate continuously.

If the forced-air heating system doesn't meet these standards, consider the following improvements.

✓ Clean or change dirty filters. Clean air-conditioning coils.

✓ Clean the blower, increase fan speed, and improve ducted air circulation. *See page 125.*

Table 5-3: Furnace Operating Parameters

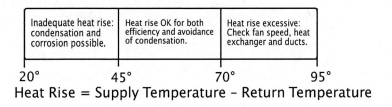

Inadequate heat rise: condensation and corrosion possible.	Heat rise OK for both efficiency and avoidance of condensation.	Heat rise excessive: Check fan speed, heat exchanger and ducts.

20°   45°   70°   95°
Heat Rise = Supply Temperature – Return Temperature

Excellent fan-off temperature if comfort is acceptable.	Borderline acceptable: Consider replacing fan control.	Unacceptable range: Significant savings possible by adjusting or replacing fan control.

85°   100°   115°   130°
Fan-off Temperature

Excellent fan-on temperature range: No change needed.	Fair: Consider fan-control replacement if fan-off temperature is also borderline.	Poor: Adjust or replace fan control.

100°   120°   140°   160°
Fan-on Temperature

✓ Adjust fan control, or replace the fan control if adjustment fails. Some fan controls aren't adjustable.

✓ Adjust the high-limit control to conform to the above standards, or replace the high-limit control.

✓ Make changes in ducts to improve airflow if necessary. *See "Duct Improvements to Increase Airflow" on page 126.*

After adjustments, measured temperature rise should be no lower than manufacturer's specifications or the listed minimum values in *Table 2-5 on page 39.*

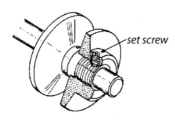

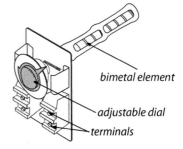

Adjustable drive pulley: This adjustable pulley moves back and forth allowing the belt to ride higher or lower, adjusting the blower's speed.

A fan/limit control: Turns the furnace blower on and off, according to temperature. Also turns the burner off if the heat exchanger gets too hot (high limit).

5.2 TROUBLESHOOTING AIRFLOW PROBLEMS

Measurements of static pressure and temperature, while they don't measure airflow (cfm) directly, are useful for troubleshooting. The tests presented in this section are merely guidelines for determining whether or not airflow is adequate.

5.2.1 Measuring Total External Static Pressure

The ducts, registers, and a coil mounted in the ducts (if present) create the duct system's resistance, which is measured by static pressure in inches of water column (IWC) or pascals. The return static is negative and the supply static is positive. Total exterior static pressure (TESP) is the sum of the absolute values of the supply and return static pressures. The positive or negative signs are disregarded when adding supply static and return static to get TESP.

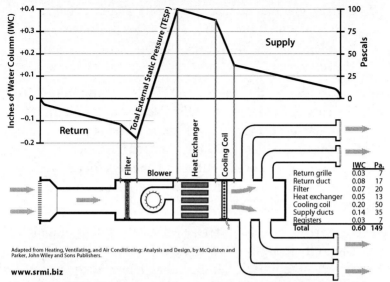

	IWC	Pa.
Return grille	0.03	7
Return duct	0.08	17
Filter	0.07	20
Heat exchanger	0.05	13
Cooling coil	0.20	50
Supply ducts	0.14	35
Registers	0.03	7
Total	0.60	149

Adapted from Heating, Ventilating, and Air Conditioning: Analysis and Design, by McQuiston and Parker, John Wiley and Sons Publishers.

www.srmi.biz

Visualizing TESP: The blower creates a suction at its inlet and a positive pressure at its outlet. As the distance between the measurement and blower increase, pressure decreases because of the system's resistance.

The greater the TESP, the lower the airflow. TESP gives a rough indicator of whether airflow is adequate. The supply and return static pressures by themselves can indicate whether the supply or return or both are restricted. For example, if the supply static pressure is 0.3 and the return static pressure is 0.7, you can assume that most of the airflow problems are due to a restricted return. The TESP test can give a very rough estimate of airflow if the manufacturer's table for static pressure versus airflow is available.

1. Attach two static pressure probes to tubes leading to the ports of the manometer. For analog manometers, attach the high-side port to the probe inserted downstream of the coil or air handler.

2. Take the readings on each side of the air handler to obtain both supply and return static pressures separately. Disregard positive or negative signs given by a digital manometer when performing addition.

3. Consult manufacturer's literature for a table, relating static pressure difference to airflow for the blower or air handler. Find airflow for the static pressure measured above.

Air handlers deliver their airflow at TESPs ranging from 0.20 IWC (75 Pascals) and 1.0 IWC (250 Pascals) as found in the field. Manufacturers maximum recommended static pressure is usually a maximum 0.50 IWC for standard air handlers. Lower TESPs are better for both comfort and energy efficiency. TESPs greater than 0.50 IWC may indicate insufficient airflow in standard residential forced-air systems.

The popularity of pleated filters, electrostatic filters, and high-static high-efficiency evaporator coils, prompted manufacturers to introduce premium air handlers that deliver adequate airflow at TESP of greater than 0.50 IWC. Premium residential air handlers can provide adequate airflow with TESP up to 0.90 IWC. TESPs greater than 0.90 IWC (225 pascals) indicate the possibility of poor airflow in these premium residential forced-air systems.

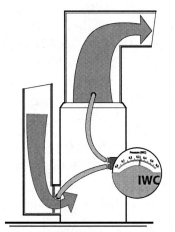

Total external static pressure (TESP): The positive and negative pressures created by the resistance of the supply and return ducts produces TESP. The measurement shown here simply adds the two static pressures without regard for their signs. As TESP increases, airflow decreases. Numbers shown below are for example only.

Total External Static Pressure Versus System Airflow

TESP (IWC)	0.3	0.4	0.5	0.6	0.7	0.8
CFM	995	945	895	840	760	670

5.2.2 Static-Pressure Drop across Coil or Filter

Measuring static pressure drop across a coil or filter can give a rough estimate of airflow when the filter or coil is new. Manufacturers often provide a table showing airflow through the filter or coil under different static pressures.

However, this test is not usually very accurate or practical. Static pressure can vary widely from point to point within the measurement area, especially when ducts take an abrupt change of direction near the air handler. Access to both sides of the coil for testing static pressure can be difficult. Drilling test holes requires care and planning to avoid damaging the indoor coil if it is located in a duct. Therefore, this test has very limited usefulness.

If you measure static pressure drop across a new or clean coil or filter, mark this reading on the unit for use as a future troubleshooting aid because the static pressure increases with dirt deposits. Also leave literature on the filter or coil with the air handler.

Static pressure drops for filters vary from 0.05 IWC to 0.20 IWC. High-static air-cleaning filters should be installed by technicians, who understand how the new filter will affect the system airflow. Indoor coils vary from 0.20 to 0.50 IWC as installed. New high-efficiency coils have higher static pressure than older coils because of smaller fin spacing. Wet coils have higher static pressure than dry coils, and the coil will likely be wet whenever the relative humidity is more than 50 percent.

The static pressure drop of a filter or coil increases as it collects dirt. However, you can't be certain that the device is dirty unless you know the device's pressure drop when the coil or filter was new or clean.

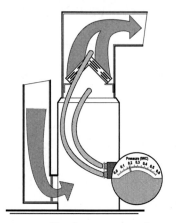

Static pressure drop across coil: The more static pressure applied to a new clean coil, the greater the airflow will be through it. Manufacturers often provide data for both dry and wet coils. Numbers listed below are for example only.

Static-pressure drop versus airflow through coil

S.P.	.23	.27	.31	.36	.41
CFM	1000	1100	1200	1300	1400

5.2.3 Unbalanced Supply-Return Airflow Test

Closing interior doors often isolates supply registers from return registers in homes with central returns. This imbalance pressurizes bedrooms and depressurizes central areas with return registers. These pressures can drive air leakage through the building shell, create moisture problems, and bring pollutants in from the crawl space or garage.

The following test uses only the air handler and a digital manometer to evaluate whether the supply air is able to cycle back through the return registers at an adequate rate. Activate the air handler and close interior doors.

✓ First, measure the changes in the pressure between the home's central living area and the outdoors with a digital manometer.

✓ Then, measure the bedrooms' pressure difference with outdoors.

If pressure difference is more than ± 2.0 pascals with the air handler operating, pressure relief is necessary. To estimate the amount of pressure relief, slowly open door until pressure difference drops to below 1 pascal. Estimate the surface area of door opening. This is the area of the permanent opening, required to provide pressure relief. Pressure relief may include undercutting the door or installing transfer grilles.

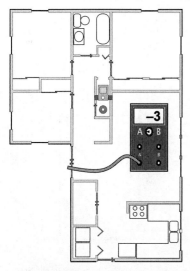

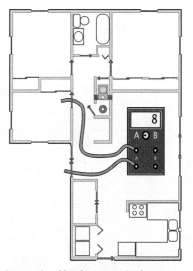

Depressurized central zone: The air handler depressurizes the central zone, where the return register is located, when the bedroom doors are closed. This significantly increases air infiltration through the building shell.

Pressurized bedrooms: Bedrooms with supply registers but no return register are pressurized when the air handler is on and the doors are closed. Pressures this high can double or triple air leakage through the building shell.

5.2.4 Carrier® Temperature-Split Airflow Evaluation

Carrier Corporation uses dry-bulb supply temperature as an indicator of whether airflow is adequate. This method requires measuring the return-air, wet-bulb and dry-bulb temperatures. From these temperatures, the recommended supply dry-bulb temperature is determined from a slide rule provided by Carrier or the table shown in *"Correct Supply Dry Bulb Temperature" on page 115*. If the measured supply dry-bulb temperature is lower than the listed value, the airflow is probably too low. If the supply dry-bulb temperature is higher than the listed value, the airflow is probably higher than 400 cfm per ton, which is usually no problem.

Carrier temperature-split method: First measure return wet-bulb and dry-bulb temperature. Check *Table 5-4 on page 115* for the suggest supply dry-bulb temperature. Then measure the actual dry-bulb supply temperature. If the measured temperature is lower than suggested, the airflow is too low.

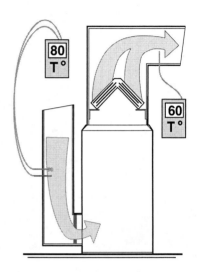

Table 5-4: Correct Supply Dry Bulb Temperature

| | Indoor Entering Air Wet Bulb F° | | | | | | | | | |
	57	59	61	63	65	67	69	71	73	75
70	51	52	53	54	55	57	59	-	-	-
72	52	53	54	55	57	58	60	62	-	-
74	53	53	55	56	58	59	61	63	65	-
76	54	54	55	57	58	60	62	64	66	-
78	55	55	56	57	59	61	63	65	67	69
80	56	56	57	58	60	62	64	66	68	69
82	57	57	58	60	61	63	65	67	69	71
84	-	59	60	61	62	63	65	67	69	71

Indoor Entering Air Dry Bulb F° (row labels)

From a calculator by Carrier Corporation, Training Division

5.2.5 Estimating Airflow of Gas Furnaces

You can estimate the airflow of a gas furnace if you have measured three key parameters.

1. Input from clocking the gas meter and knowing the gas utility's specific heat content per cubic foot of gas.

2. Steady-state efficiency (SSE) from a modern combustion analyzer.

3. Temperature rise across between the supply and return plenums.

Divide the input by the SSE to get the output. Then use *Table 5-5 on page 116* to find the approximate airflow.

Table 5-5: Approximate Airflow in CFM from Output and Temperature Rise

Output kBTUH	Temperature Rise F°						
	30	40	50	60	70	80	90
40	1250	950	750	600	550	450	400
50	1550	1150	950	750	650	600	500
60	1850	1400	1100	950	800	700	600
70	2150	1600	1300	1100	950	800	700
80	2450	1850	1500	1250	1050	950	800
90	2800	2100	1650	1400	1200	1050	950
100	3100	2300	1850	1550	1300	1150	1050
115	–	2650	2150	1800	1550	1350	1200
130	–	3000	2400	2000	1700	1500	1350
145	–	–	2700	2250	1900	1700	1500
160	–	–	2950	2450	2100	1900	1650
175	–	–	–	2700	2300	2050	1800

CFM = output kBTUH / 1.08 x ⊠T; Rounded to 2.5 significant digits.

Air-handler airflow greater than 15cfm/kBTUH is ideal for condensing furnaces. This level of airflow allows combustion gasses to cool in the heat exchanger, which promotes condensation and high efficiency in condensing furnaces. Airflows significantly less than 15cfm/kBTUH may result in less condensation and lower efficiencies.

Low- and mid-efficiency furnaces operating at greater than 82% efficiency or less than 300°F flue-gas temperature are at risk of excessive flue-gas condensation resulting in corroded heat-exchangers and flues. Therefore the airflow in these units may need to be less than 15cfm/kBTUH to ensure adequate flue-gas temperatures to power the venting system.

5.2.6 Measuring Supply-Register Airflows

Flow hoods are designed to measure airflow through individual supply or return registers. How accurately flow hoods measure these airflows, especially airflows less than 75 CFM, is subject to debate.

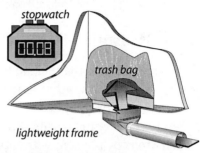

stopwatch

trash bag

lightweight frame

If you have a flow hood, follow the manufacturer's instructions for calibrating and using it to measure supply airflows. If you don't have access to a flow hood, a large plastic trash bag and a stopwatch can measure supply-register airflow

$$33 \text{ GAL} \times 0.134\frac{\text{FT}^3}{\text{GAL}} = 4.42 \text{ FT}^3$$

$$\frac{8 \text{ SECONDS}}{60 \text{ SECONDS}} = 0.13 \text{ MINUTES}$$

$$\frac{4.42 \text{ FT}^3}{0.13 \text{ MINUTES}} = 34 \text{ CFM}$$

fairly accurately. The trash bag is attached to a lightweight frame, such as a large duct boot, take-off, or plastic container with the bottom cut out. This frame helps it to cover the register and capture all its airflow. The stopwatch measures the amount of time it takes to fill the bag.

✓ Collapse the trash bag and position it over the register as you start the stopwatch.

✓ Note the time required to fill the bag.

✓ Take the volume of the bag in cubic feet (not gallons) and divide by the time in minutes to get CFM, as shown in the illustration.

5.3 MEASURING AIRFLOW

The advantage of measuring airflow is that airflow is directly related to seasonal efficiency. If measured airflow is incorrect, you troubleshoot the problem, make improvements, and then measure the subsequent change in airflow.

5.3.1 Measuring Return Airflow with a Flow Hood

This test measures the relatively laminar airflow at the single or few return registers rather than the relatively turbulent airflow at the many supply registers. The flow-hood inlet should be larger than the return grills, although 10 percent of the return register may be blocked with tape to allow the flow hood to cover that reduced opening.

545 + 495 = 1140 cfm

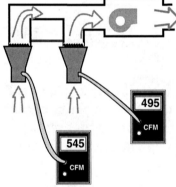

Measuring return air with a flow hood: This method provides an estimate of system airflow. It can significantly underestimate airflow because return duct leakage bypasses the measurement.

This flow-hood test works best on systems where the return ducts aren't too leaky. The return airflow will appear low if the return ducts are leaky. A low reading means that the airflow is restricted or that the system is drawing considerable return air from a crawl space, attic, or attached garage.

One or more return grills should be located in areas where a flow hood can cover the grills and be centered over them. Turn on the air handler to run at the higher fan speed, normally used for cooling.

1. Center the flow hood over the return register, covering it completely. If the register is larger than the flow hood,

seal up to 10 percent of the register with tape before covering it.

2. Read and record the airflows through the return registers. Add the measured airflows of the return registers together to get the total system airflow.

This method isn't usually as accurate as measuring with a duct blower or flow plate, discussed next. The flow plate measures airflow right at the air handler, not at the return. And, the duct-blower airflow test simply duplicates the normal airflow, taking the leaky return ducts out of the loop by blocking them off.

5.3.2 Flow-Plate Method for Measuring System Airflow

The TrueFlow® air-handler flow meter, manufactured by The Energy Conservatory, is relatively fast and easy to use compared to the duct blower, discussed next. The TrueFlow® meter is a plate with holes and sampling tubes that samples and averages pressures and converts them into an airflow measurement.

One of two flow plates are inserted and sealed at their edges in the filter slot or bracket within the air handler. Then, the flow plate measures air-handler airflow, using a specially programmed digital manometer.

When used according to the manufacturer's instructions, which are summarized below, the accuracy of this method is better than the other tests described on these pages, with the exception of the duct blower test. Refer to the manufacturer's instructions for the precise testing methodology. A summary of flow-plate procedures follows.

1. Measure and record the normal system operating pressure, with a standard filter in place, using a static pressure probe in the supply plenum or supply duct near the air handler.

2. Replace the existing filter with the flow plate. Seal the flow plate into the slot, according to the manufacturer's recommendations.

3. Measure and record the system's operating pressure with the flow plate in place, at the same location as when the filter was in place.

4. Measure the flow through the TrueFlow Meter using the digital manometer supplied by the manufacturer.

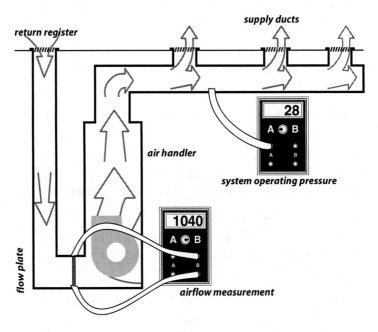

True Flow® Meter: The True Flow® flow plate installs in a filter slot and measures system airflow almost as accurately as the duct blower.

5.3.3 Duct-Blower Airflow Measurement

The duct blower is a fan mounted in an aerodynamic housing and equipped with a pressure-sampling tube. The duct blower is the most accurate airflow-measuring device currently available.

During this airflow test, all return air is routed through the duct blower where the airflow can be measured. The return air traveling through the duct blower is moved by the air handler's blower aided by the duct blower.

Main return is disconnected and the air handler is sealed.

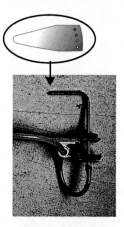

Duct blower mounted to air handler: The best way to measure airflow with a duct blower is to connect the duct blower to the air handler and seal off the main return.

Static pressure probe: Tiny holes near the tip sense static pressure.

1. Set up a static pressure gauge to measure the duct pressure in the supply plenum, or a few feet away from the supply plenum, in a main supply. Tape the static pressure probe to hold it in place. The point of the probe should face into the oncoming airflow with the probe's static-pressure openings perpendicular to the airflow direction.

2. Make sure all supply registers and return grills are open. Leave filters installed.

3. Turn on the system and measure static pressure from the probe installed in Step 1.

4. Shut off power to the air handler. Connect the duct blower to blow into the air handler at the blower compartment or into the single return register, either by blocking the main return or connecting the duct blower to the main return.

Pressure in two measurement systems: Technicians and engineers use both Pascals (metric) and inches of water column (American) to measure duct pressures.

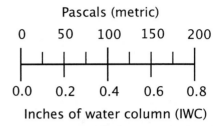

Blocking the Main Return

The preferred method of connecting the duct blower is to block the return plenum's main return entry to the air handler in one of the following ways.

- Install the temporary barrier in a filter slot.

- Support the main return and move it temporarily out of your way, while you seal the opening to the air handler with cardboard and tape.

- If there is room, install a cardboard barrier from the open door of the blower compartment, taping the edges of the perimeter to air-seal the barrier. Be careful not to scratch your arms and hands.

After installing and sealing the barrier, connect the duct blower to the blower compartment after removing the door.

Connecting the Duct Blower to a Main Return

Remove the grill at the single return register. Connect the duct blower through its flexible tube or else directly to the register, using cardboard to block off the excess area of the register. (Note: If there is significant return leakage, airflow measurement will be artificially high.)

All the return air should now flow through the duct blower. If the duct blower is connected to an air handler, located outside the conditioned space, the door or access panel between the conditioned space and the air handler location must be opened. Now you are ready to measure system airflow.

Duct blower mounted to main return: With a single return, it's convenient to attach the duct blower to the single main return register. However, this option may result in an artificially high airflow reading.

1. Turn on the air-handler fan once again, making sure the air-handler fan is running at the correct speed for cooling.

2. Turn on the duct blower to blow into the air handler, increasing airflow until the manometer measuring supply-plenum static pressure reads the same as your original static-pressure measurement.

3. Measure and record the airflow through the duct blower. Refer to the duct-blower instruction book, if necessary, to insure that you know how to take the reading. The airflow reading you take directly from the digital manometer or look up in the manufacturer's

table for converting pressure to flow is total system air-flow in cubic feet per minute (CFM).

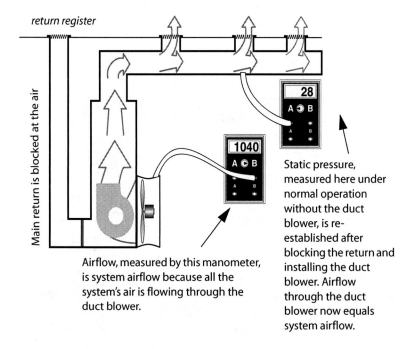

return register

Main return is blocked at the air

28
A ○ B

1040
A ○ B

Airflow, measured by this manometer, is system airflow because all the system's air is flowing through the duct blower.

Static pressure, measured here under normal operation without the duct blower, is re-established after blocking the return and installing the duct blower. Airflow through the duct blower now equals system airflow.

5.4 IMPROVING DUCT AIRFLOW

Inadequate airflow is a common cause of comfort complaints. When the air handler is on there should be a strong flow of air out of each supply register. Low airflow may mean that a branch is blocked or separated, or that return air is not sufficient. When low airflow is a problem, consider the following obvious improvements mentioned previously.

✓Clean or change filter.

✓Clean furnace blower.

✓Clean air-conditioning or heat pump coil. (If the blower is dirty, the coil is probably also dirty.)

✓Increase blower speed.

✓ Make sure that balancing dampers to rooms that need more airflow are wide open.

✓ Lubricate blower motor, and check tension on drive belt.

✓ Repair or replace bent, damaged, or restricted registers.

Washable filter installed on a rack inside the blower compartment.

Panel filter installed in filter slot in return plenum.

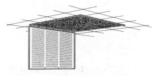

Panel filter installed in return register.

Furnace filter location: Filters are installed on the return-air side of forced air systems. Look for them in one or more of the above locations.

5.4.1 Filter and Blower Maintenance

A dirty filter can reduce airflow significantly. Take action to prevent filter-caused airflow restriction by taking the following steps.

✓ Make sure that the filter fills entire airway downstream of the blower, heat exchanger, and air-conditioning coil. The filter should fit tightly against its frame or the surface of the air handler or duct to prevent airflow around the filter.

✓ Insure that filters are easy to change or clean.

✓ Stress to the client the importance of changing or cleaning filters, and suggest to the client a regular filter-maintenance schedule.

✓ Clean the blower if it is dirty. This task involves removing the blower and removing dirt completely with a brush or water spray.

✓ Special air-cleaning filters offer more resistance than standard filters, especially when saturated with dust. Avoid using them, unless you test for airflow after installation.

✓ Measure the current draw of the blower motor in amps. If the amp measurement exceeds the motor full load amp (FLA) rating by more than 10%, replace the motor.

5.4.2 Duct Improvements to Increase Airflow

Consider the following duct changes to increase system airflow and reduce the imbalance between supply and return ducts.

✓ Clean dirty filters and modify the filter installation to allow easier filter changing, if this is difficult.

✓ Remove obstructions to registers and ducts such as rugs, furniture, and objects placed inside ducts, such as children's toys and water pans for humidification.

✓ Remove kinks from flex duct, and replace collapsed flex duct and fiberglass duct board.

✓ If the blower is dirty, an air conditioning coil, if present, is probably also dirty. Clean the blower and coil as outlined in *"Cleaning Blowers and Indoor Coils" on page 152.*

✓ Install additional supply ducts and return ducts as needed to provide heated air throughout the building, especially in additions to the building.

✓ Install a transfer grill between the bedroom and main body of house to improve airflow.

✓ Undercut bedroom doors, especially in homes with single return registers.

✓ Retrofit jumper ducts, composed of one register in the bedroom, one register in the central return-air zone, and a duct in between (usually running through an attic or crawl space).

✓Install registers and grills where missing.

5.4.3 Adding New Ducts

New ducts should not be installed in unconditioned spaces unless absolutely necessary. If ducts are located in unconditioned spaces, joints should be sealed and the ducts insulated as described in this guide. *See pages 130 and 142.*

New ducts must be physically connected to the existing distribution system or to the furnace. Install balancing damper in each new branch duct. Registers should terminate each new supply or return branch duct.

Anthony Cox

Adding a return duct: Return air is often inadequate in existing forced-air distribution systems.

The return side of the forced-air system is often the most restricted side and the one needing an increased volume of air from the house. It 's nearly impossible to provide too many square inches of cross-sectional area of return ducts.

Connecting a new duct to the existing return main duct won't necessarily improve airflow because this approach doesn't increase the cross-sectional area of main duct connected to the blower compartment. A standard air handler usually has three or four sides where a large return duct can attach. If the blower is very close to one of the air handler's walls, that side should not be used to attach another return duct. Choose a central location as far away as possible from the existing return register to install a new return register. Run flexduct or metal duct from the air handler to the new return.

When a main metal duct trunk makes a 90° bend, usually near the air handler, retrofit turning vanes may increase airflow significantly.

Jumper returns or pass-through grills are also effective to increase airflow as long as you don't create air leaks in existing ducts or the building shell by installing them.

When a return register is installed in a floor or ceiling, you can enlarge the return by adding a large sheet-metal box with a filter inside with space for installing one or two additional returns. Whenever the air handler is accessible, install the filter in or near the air handler so that any duct leakage will be filtered. Return leaks can deposit a lot of dust on heat exchangers, blowers, and coils.

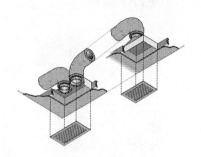

Jumper ducts can bring air from a restricted area of the home back to a main return register.

Installing transfer grills in doors or through walls allows return air to escape from bedrooms.

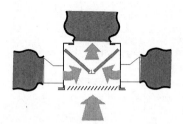

You can expand an existing return register by connecting a return box and then run one or more branch returns into the box in a way that all the air is filtered.

Turning vanes dramatically improve airflow in square-duct bends.

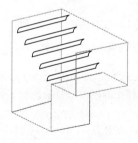

Restricted return air: Return air is often restricted, requiring a variety of strategies to relieve the resulting house pressures and low system airflow. Installing an additional return duct directly into the air handler is a preferred strategy.

5.5 MEASURING DUCT AIR LEAKAGE

Duct air leakage is a major energy-waster in homes where the ducts are located outside the home's thermal boundary in a crawl space, attic, attached garage, or leaky unoccupied basement. When these intermediate zones will remain outside the thermal boundary, duct air-sealing is often cost-effective.

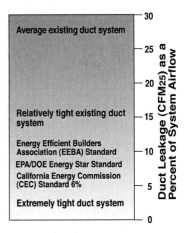

Duct leakage standards: Dividing the tested duct leakage in CFM_{25} by the system airflow provides the most common type of duct-leakage standard.

Ducts should be tested to determine how much they leak before any duct air sealing is performed.

5.5.1 Duct Air-Tightness Standards

Duct leakage should always be minimized by sealing ducts during installation. Unfortunately, this simple standard has been ignored in most homes. A study commissioned by the EPA, quoted earlier, states that duct leakage averages around 300 CFM_{25} among 19 duct-leakage studies, in 14,000 homes. This average duct leakage represents 25 to 35 percent of assumed airflow through the air handler.

Duct leakage may or may not be a significant energy and comfort problem, depending on where the ducts are located. If the ducts are located completely within the conditioned living space of a home, duct leakage is probably not a significant energy problem. An example of ducts completely within the living space is a two-story home with 1000 square feet of area on each floor where the entire floor area is heated and cooled.

If existing ducts are located outside the home's thermal boundary, it's a safe bet that duct-sealing is cost-effective, assuming you don't have to demolish anything. You can either assume high typical air leakage and begin sealing the system's leaks, working from the air handler out to the extremities, using touch and sight to find the leaks. Or you can measure duct leakage and decide how much labor and materials is merited. Duct-leakage testing also helps measure success and gives technicians valuable feedback.

Table 5-6: Total Duct Air Leakage Standards for Homes

	CFM_{50}	CFM_{25}
Existing Homes	10% of floor area	6% of floor area
New Homes	6% of floor area	3.5% of floor area

From Delta-T Inc. and Oregon Department of Energy

5.5.2 Measuring Duct Air Leakage with a Duct Blower

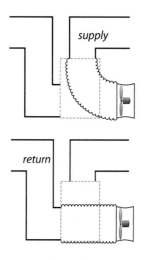

supply

return

Testing ducts before air-handler installation: If the ducts are installed prior to the air handler, as with a furnace replacement or many new installations, the duct blower can test first supply then return ducts for airtightness.

Pressurizing the ducts with a duct blower measures total duct leakage. The duct blower is the most accurate common measuring device for duct air leakage. It consists of a fan, a digital manometer or set of analog manometers, and a set of reducer plates for measuring different leakage levels. Using a blower door with a duct blower measures leakage to outdoors.

Measuring Total Duct Leakage

The total duct leakage test measures leakage to both indoors and outdoors. The house and intermediate zones should be open to the outdoors by way of windows, doors, or vents. Opening the intermediate zones to outdoors insures that the duct blower is measuring only the ducts' airtightness — not the airtightness of ducts combined with other air barriers like roofs, foundation walls, or garages.

Supply and return ducts can be tested separately, either before the air handler is installed in a new home or when an air handler is removed during replacement.

Follow these steps when performing a duct airtightness test.

1. Install the duct blower in the air handler or to a large return register, either using its connector duct or simply attaching the duct blower itself to the air handler or return register with cardboard and tape.

2. Remove the air filter(s) from the duct system.

3. Seal all supply and return registers with masking tape or other non-destructive sealant.

4. Open the house, basement or crawl space, containing ducts, to outdoors.

5. Drill a $^1/_4$ or $^5/_{16}$-inch hole into a supply duct a short distance away from the air handler and insert a manometer hose. Connect a manometer to this hose to measure *duct WRT outdoors*. (Indoors, outdoors, and intermediate zones should ideally be opened to each other in this test).

6. Connect an airflow manometer to measure *fan WRT the area near the fan*.

Check manometer(s) for proper settings. Digital manometers require your choosing the correct mode, range, and fan-type settings.

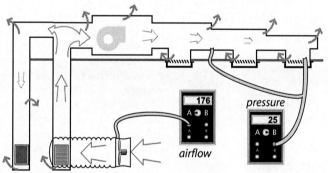

Total duct air leakage measured by the duct blower: All registers are sealed except the one connecting the duct blower to the system. Pressurize the ducts to 25 pascals and measure airflow.

1. Turn on the duct blower and pressurize the ducts to 25 pascals.

2. Record duct-blower airflow.

3. While the ducts are pressurized, start at the air handler and move outward feeling for leaks in the air handler, main ducts, and branches.

4. After testing and associated air-sealing are complete, restore filter(s), remove seals from registers, and check air handler.

Measuring Duct Leakage to Outdoors

Measuring duct leakage to outdoors gives you a duct-air-leakage value that is directly related to energy waste and the potential for energy savings.

1. Set up the home in its typical heating and cooling mode with windows and outside doors closed. Open all indoor conditioned areas to one another.

2. Install a blower door, configured to pressurize the home.

3. Connect the duct blower to the air handler or to a main return duct.

Measuring duct leakage to outdoors: Using a blower door to pressurize the house with a duct blower to pressurize the ducts measures leakage to the outdoors—a smaller number and a better predictor of energy savings. This test is the preferred for evaluating duct leakage for specialists in both shell air leakage and duct air leakage whenever a blower door is available.

4. Pressurize the ducts to +25 pascals by increasing the duct blower's speed until this value is reached.

5. Pressurize the house until the pressure difference between the house and duct is 0 pascals (*house WRT ducts*).

6. Read the airflow through the duct blower. This value is duct leakage to outdoors.

5.6 TROUBLESHOOTING DUCT LEAKAGE

After you know that duct air leakage is a problem, there are several methods for finding the locations of the leaks and evaluating their severity.

5.6.1 Finding Duct Leaks Using Touch and Sight

One of the simplest ways of finding duct leaks is feeling with your hand for air leaking out of supply ducts, while the ducts are pressurized by the air handler's blower. Duct leaks can also be located using light. Here are three different tests used to locate air leaks.

1. Use the air handler blower to pressurize supply ducts. Closing the dampers on supply registers temporarily or partially blocking the register with pieces of carpet, magazines, or any object that won't be blown off by the register's airflow will increase the duct pressure and make duct leaks easier to find.

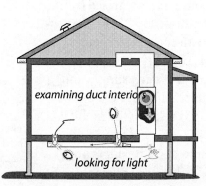

examining duct interior

looking for light

feeling for air

Finding duct air leaks: Finding the exact location of duct leaks precedes duct air-sealing.

2. Place a trouble light, with a 100-watt bulb, inside the duct through a register. Look for light emanating from the exterior of duct joints and seams.

3. Recognize which duct joints were difficult to fasten and seal during installation. These joints are likely duct-leakage locations.

Feeling air leaks establishes their exact location. Ducts must be pressurized in order to feel leaks because you can't usually feel air leaking into depressurized ducts.

A trouble light, flashlight, and mirror help you to visually understand duct interiors so that you can plan an air-sealing procedure.

5.6.2 Pressure-Pan Testing

Pressure-pan tests can help identify leaky or disconnected ducts. The pressure-pan test requires a blower door. With the house depressurized by the blower door to –25 or –50 pascals with reference to outdoors, pressure-pan readings are taken at each supply and return register. Pressure-pan testing is reliable for mobile homes and small site-built homes where the ducts are outside the air barrier.

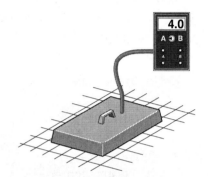

A pressure pan: Blocks a single register and measures the air pressure behind it, during a blower door test. The magnitude of that pressure is an indicator of duct leakage.

1. Install blower door and set-up house for winter conditions. Open all interior doors.

2. If the basement is conditioned living space, open the door between basement and upstairs living spaces. If the basement is considered outside the conditioned living space, close the door between basement and upstairs living spaces and open a basement window.

3. Turn furnace off. Remove furnace filter, and tape filter slot if one exists. Ensure that all grills, registers, and dampers are fully open.

4. Temporarily seal any outside fresh-air intakes to the duct system. Seal supply registers in zones that are not

intended to be heated—an uninhabited basement or crawl space, for example.

5. Open attics, crawl spaces, and garages as much as possible to the outside so they don't create a secondary air barrier.

6. Connect hose between pressure pan and the input tap on the digital manometer. Leave the reference tap open.

7. With the blower door at –25 pascals, place the pressure pan completely over a grill or register to form a tight seal. Record the reading, which should be a positive pressure.

8. If a grill is too large or a supply register is difficult to access (under a kitchen cabinet, for example), seal the grill or register with masking tape. Insert a pressure probe through the masking tape and record reading.

9. Repeat this test for each register and grill in a systematic fashion.

Pressure-Pan Duct Standards

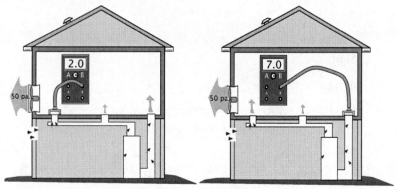

Pressure-pan test: A pressure-pan reading of 2 indicates moderate duct air leakage.

Problem register: A pressure reading of 7 pascals indicates major air leakage near the tested register.

If the ducts are perfectly sealed with no leakage to the outside, no pressure difference (0 pascals) will be measured during a pressure-pan test. The higher the measured pressure-pan reading, the more connected the duct is to the outdoors. Readings greater than 1.0 pascal require investigation and sealing of leaks causing the reading.

Pay particular attention to registers connected to ducts that are located in areas outside the conditioned living space. These spaces include attics, crawl spaces, garages, and unoccupied basements. Also test return registers attached to stud cavities or panned joists used as return ducts. Leaky ducts located outside the conditioned living space may show pressure-pan readings of up to 25-to-50 pascals if they have large leaks.

5.6.3 Measuring House Pressure caused by Duct Leakage

The following test measures pressure differences between the main body of the house and outdoors, caused by duct leakage. Pressure difference greater than +2.0 pascals or more negative than −2.0 pascals should be corrected.

1. Set-up house for winter conditions. Close all windows and exterior doors. Turn-off all exhaust fans.

2. First, open all interior doors, including door to basement.

3. Turn on air handler.

4. Measure the house-to-outdoors pressure difference. This test indicates dominant duct leakage as shown here.

A positive pressure indicates that the return ducts (which pull air from leaky intermediate zones) are leakier than the supply ducts. A negative pressure indicates that the supply ducts (which push air into intermediate zones through their leaks) are leakier than return ducts. A pressure at or near zero indicates equal supply and return leakage or else little duct leakage.

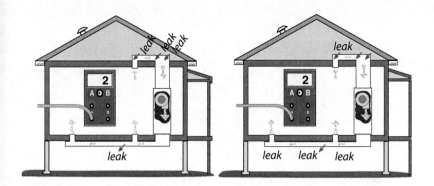

Dominant return leaks: When return leaks are larger than supply leaks, the house shows a positive pressure with reference to the outdoors.

Dominant supply leaks: When supply leaks are larger than return leaks, the house shows a negative pressure with reference to the outdoors.

5.7 Sealing Duct Leaks

Ducts located outside the thermal boundary or in an intermediate zone like a ventilated attic or crawl space should be sealed. The following is a list of duct-leak locations in order of their relative importance. Leaks nearer to the air handler see higher pressure and are more important than leaks further away.

✓ First, seal all return leaks within the combustion zone to prevent this leakage from depressurizing the combustion zone and causing backdrafting.

✓ Plenum joint at air handler: These joints may have been difficult to fasten and seal because of tight access. Seal these thoroughly even if it requires cutting an access hole in the plenum. (Prefer silicone caulking or foil tape to mastic and fabric mesh here for future access—furnace replacement, for example.)

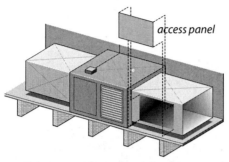

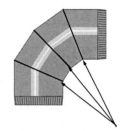

Plenums, poorly sealed to air handler: When air handlers are installed in tight spaces, plenums may be poorly fastened and sealed. Cutting a hole in the duct may be the only way to seal this important joint.

Sectioned elbows: Joints in sectioned elbows known as gores are usually quite leaky and require sealing with duct mastic.

✓Joints at branch takeoffs: These important joints should be sealed with a thick layer of mastic. Fabric mesh tape is a plus for new installations or when access is easy.

✓Joints in sectioned elbows: Known as gores, these are usually leaky.

✓Tabbed sleeves: Attach the sleeve to the main duct with 3-to-5 screws and apply mastic plentifully.

✓Flexduct-to-metal joints: Clamp the flexduct's inner liner with a plastic strap, using a strap tensioner. Clamp the insulation and outer liner with another strap.

✓Support ducts and duct joints with duct hangers where needed. Install duct hangers approximately every 24 inches.

✓Seal leaky joints between building materials composing cavity-return ducts, like panned floor cavities and furnace return platforms.

✓Deteriorating ductboard facing: Replace ductboard, preferably with metal ducting when the facing deteriorates because this condition leads to massive air leakage.

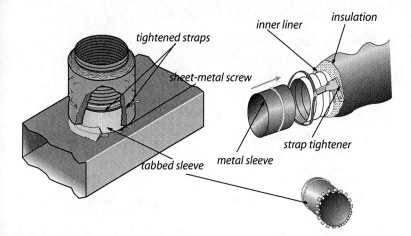

tightened straps

sheet-metal screw

inner liner

insulation

strap tightener

tabbed sleeve   metal sleeve

Flexduct joints: Flexduct itself is usually fairly airtight, but joints, sealed improperly with tape, can be very leaky. Use methods shown here to make flexduct joints airtight.

✓ Seal leaky joints between supply and return registers and the floor, wall, and ceiling to which they are attached.

✓ Consider sealing off supply and return registers in unoccupied basements or crawl spaces.

✓ Seal penetrations made by wires or pipes traveling through ducts. Or better: move the pipes and wires and patch the holes.

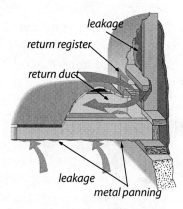

leakage

return register

return duct

leakage

metal panning

Panned floor joists: These return ducts are often very leaky and may require removing the panning to seal the cavity.

5.7.1 Materials for Duct Air-Sealing

Duct mastic is the preferred duct-sealing material because of its superior durability and adhesion. Apply at least $1/16$-inch thick and use reinforcing mesh for all joints wider than $1/8$ inch or joints that may experience some movement.

Siliconized acrylic-latex caulk is acceptable for sealing joints in panned joist spaces, used for return ducts.

Duct mastic: Mastic, reinforced with fabric webbing, is the best choice for sealing ducts.

Joints should rely on mechanical fasteners to prevent joint movement or separation. Tape should never be expected to hold a joint together nor expected to resist the force of compacted insulation or joint movement. Aluminum foil or cloth duct tape are not good materials for duct sealing because their adhesive often fails after a short time.

5.8 DUCT INSULATION

Insulate supply ducts that run through unconditioned areas outside the thermal boundary such as crawl spaces, attics, and attached garages with a minimum of R-6 vinyl- or foil-faced duct insulation. Don't insulate ducts that run through conditioned areas unless they cause overheating in winter or condensation in summer. Follow the best practices listed below for installing insulation.

✓Always perform necessary duct sealing before insulating ducts.

✓ Insulation should cover all exposed supply ducts, without significant areas of bare duct left uninsulated.

✓ Fasten insulation using mechanical means such as stick pins, twine, or plastic straps. Tape can be effective for covering joints in the insulation to prevent air convection, but tape usually fails if expected to resist the force of the insulation's compression or weight.

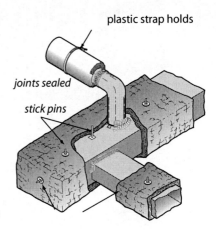

plastic strap holds

joints sealed

stick pins

duct insulation fastened

Duct insulation: Supply ducts, located in unheated areas, should be insulated to a minimum of R-6.

CHAPTER 6: COOLING AND HEAT PUMPS

Mechanical cooling systems include air conditioners and evaporative coolers. Each of these can be either be a room-cooling unit or a central unit with ducts. Heat pumps are reversible air conditioners, providing winter heating in addition to summer cooling.

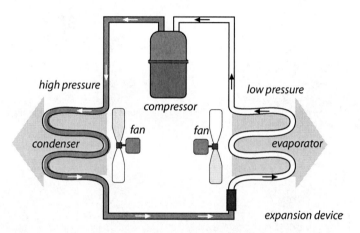

Refrigeration and air-conditioning cycle: Refrigerant evaporates in the evaporator, absorbing heat from the metal coil and the indoor air. The compressor compresses the refrigerant, and moves it to the condenser. In the condenser, the condensing refrigerant heats the condenser coil to a significantly higher temperature than the outdoor air. The outdoor air that circulates through the condenser, removing the heat that was collected from indoor air.

Air conditioners come in two basic types, packaged systems and split systems. Packaged air conditioners include room air conditioners and room heat pumps, along with packaged central air conditioners and heat pumps mounted on roofs and on concrete slabs outdoors. Split systems have an outdoor coil packaged with the compressor and controls and an indoor coil, located inside a furnace, heat pump, or main supply duct. Mini-split systems have a condenser outdoors and one or more evaporators in rooms.

Room air conditioners and heat pumps and mini-split central air-conditioning and heat-pump systems are inherently more efficient than ducted central systems because they don't use ducts, which often waste 30% of the system's energy.

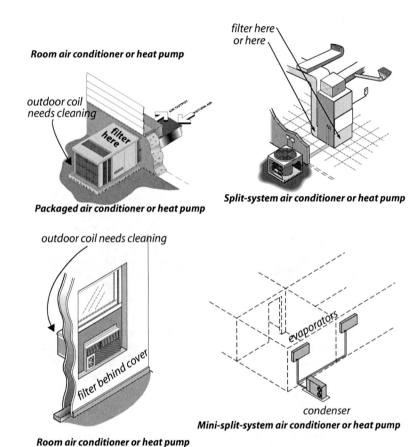

Air conditioner types: All types of air conditioners need clean filters and coils to achieve acceptable efficiency.

6.1 REPLACING AIR CONDITIONERS AND HEAT PUMPS

Replacing older inefficient air conditioners and heat pumps with new efficient units can be a cost-effective energy conservation measure. In many homes, replacing units involves only swapping boxes. In other homes, the ducts need extensive repair and re-configuring. And in some homes, it may be wise to discard an old central ducted forced-air system in favor of a more efficient mini-split or room unit.

When replacing room air conditioners or room heat pumps, simply choose the most efficient model. Room units offer very efficient heating and cooling.

Mini-split air conditioners and heat pumps offer the most efficient heating and cooling for warm and moderate climates. Mini-splits also offer flexibility for buildings without ducts or with duct systems that are obsolete, worn out, or otherwise inappropriate.

6.1.1 Replacing Forced-Air Systems

Central forced-air heat-pump and air-conditioning systems present two choices for replacement. The first choice is to make appropriate modifications to the existing duct system to optimize performance. The second choice is to discard the ducts and replace the forced-air system with a ductless mini-split system.

If the existing ducts will remain, evaluate the existing forced-air system to identify duct modifications necessary to optimize performance and efficiency. Duct modifications include repair, sealing, installation of additional branch ducts, or complete duct replacement. See "Evaluating Forced-Air System Performance" on page 101.

Evaluating and Repairing the Existing Ducts

Replacement of a forced-air furnace, air conditioner, or heat pump should include all necessary duct modifications, electrical repairs, and other improvements necessary to create a "like new" ducted space-conditioning system.

Replacing a Ducted System with A Ductless System

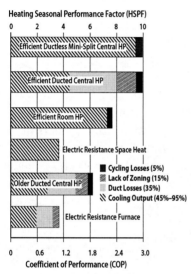

Comparison of approximate electrical heating efficiencies:

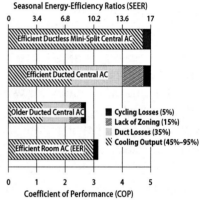

Comparison of approximate cooling efficiencies:

6.1.2 Choosing Air Conditioners and Heat Pumps

6.2 Cleaning Air-Conditioning Coils

Clean filters and air-conditioning coils are a minimum requirement for efficient operation. For more information on filters and their location, see *page 125*.

Keeping filters clean is the best way to keep coils clean. Cleaning an indoor air-conditioning coil is much more difficult than changing or cleaning a filter. When a filter is dirty or absent, dirt collects on the coil, fan blades, and other objects in the air stream. The dirt deposits reduce airflow and will eventually cause the air-conditioning system to fail.

Dirt builds up on a coil from the side where the air enters. The heaviest deposits of lint, hair, and grease coat that side of the indoor coil.

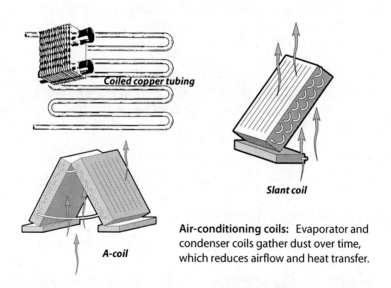

Coiled copper tubing

Slant coil

A-coil

Air-conditioning coils: Evaporator and condenser coils gather dust over time, which reduces airflow and heat transfer.

The outdoor coils of air conditioning systems aren't protected by filters. They get dirty depending on how much dust is in the outdoor air. If there is little dust and pollen in an area, the outdoor coil may only need cleaning every three years or so. However, if there is a lot of pollen and dust, annual cleaning is a good practice.

Coil-Cleaning Methods

If the blower of an air handler has dirty blower blades, the indoor coil is probably also dirty. It is a safe assumption that all outdoor coils need cleaning.

A garden chemical sprayer is a good tool for washing dirt out of indoor coils because it produces a fine air-powered spray without too much water volume. Delicate adjustable pressure washers are also available with a variety of nozzles. Consider the following methods for cleaning outdoor coils.

- If you see heavy dirt on the outside of the coil, rake the dirt off with a stiff hair brush or other suitable brush. Wetting the dirt often helps to soften and loosen it.

- With a garden sprayer, spray pressurized cold water and coil cleaner from the opposite direction of airflow to push the dirt out the same way it came in.

- Apply water with or without coil cleaner with a low-volume adjustable pressure washer, designed for coil cleaning. Apply the pressurized water at an angle and be careful not to bend fins.

- After the coil is clean, use a fin comb to straighten any bent fins.

6.2.1 Cleaning Room A/C and Heat-Pump Coils

Room air conditioners have foam or fiberglass filters that lie up against the indoor coil. It's good practice to carry a roll of filter material to replace worn-out or missing filters. Cleaning the indoor coil is easy since the heaviest dirt collects on the surface of the coil facing the inside of the home. Cleaning the outdoor coil is more difficult. Usually cleaning the outdoor coil involves removing the room air conditioner from the window and taking it to an outdoor location where you can use a garden sprayer or adjustable pressure washer.

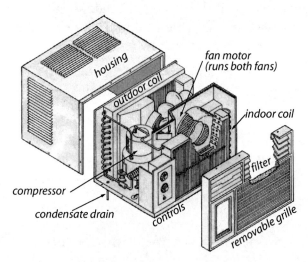

Cleaning room air conditioners: Room-air-conditioner performance deteriorates as it accumulates dirt. The unit will eventually fail to cool the room or break down unless cleaned.

Observe the following steps when cleaning the indoor and outdoor coils of a room air conditioner.

1. Remove the grille and filter on the interior side of the unit.

2. Unplug and remove the air conditioner temporarily from the window or wall. With some units, the mechanical parts slide out of the housing, and with others you must remove the whole unit, housing and all.

3. Take the unit to an clean outdoor area that drains well, like a driveway or patio.

4. Cover the compressor, fan motor, and electrical components with plastic bags, held in place with rubber bands.

5. Dampen each of the coils with a light spray of water, then rake as much dirt off the coils as you can with an old hairbrush.

6. Spray indoor coil cleaner into the indoor coil and outdoor coil cleaner into the outdoor coil, and let the cleaner set for a minute or two.

7. Rinse the cleaner and dirt out of the coils with a gentle spray from a garden sprayer or adjustable pressure washer.

8. Repeat the process again until the water draining from the coils is clean.

9. Straighten bent fins with a fin comb to prevent bent fins from reducing airflow.

6.2.2 Cleaning Blowers and Indoor Coils

Every indoor coil should be protected by an air filter that fills the entire cross-sectional area of return duct leading to the blower and indoor coil. Filters are easier to change or clean compared to cleaning a blower or coil. When equipped with clean well-fitting filters, the blower and coil remain clean for many years. However, many coils haven't had the benefit of clean, well-fitting filters and are packed with dirt. Consider the following steps to cleaning blowers and indoor coils on central air-conditioning systems.

1. Shut off the main switch to the air handler.

2. Open the blower compartment and look into the blades of the blower, using a flashlight. Reach in and slide your finger along a fan blade. Have you collected a mound of dust?

3. If the blower is dirty, remove it and clean it. If you remove the motor, you can use hot water or household cleanser to remove the dirt.

4. If the blower is dirty, the indoor coil is probably also dirty. Inspect the coil visually if you have access. Create an access hatch if needed.

5. If the coil is dirty, clean it using a brush, indoor coil cleaner and sprayed water.

6. Straighten bent fins with a fin comb to prevent bent fins from reducing airflow.

7. Clean the drain pan and drain line.

8. Vacuum dirt and water from air handler components that were dirtied by this cleaning process

A wet vacuum cleaner helps greatly in cleaning up water and dirt from the coil during this process.

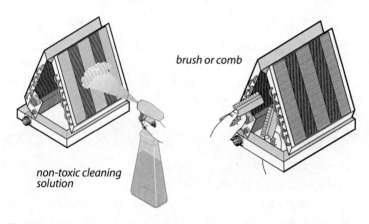

brush or comb

non-toxic cleaning solution

Cleaning an A-coil: A-coils are found in upflow and downflow air handlers. In downflow models the dirt collects on top and on upflow units dirt collects on the bottom. Clean the coil, drain pan, and drain line.

6.3 Evaluating Refrigerant Charge

The performance and efficiency of residential and light-commercial air conditioners and heat pumps is very dependent upon having the correct amount of refrigerant in the system. This section describes accepted procedures for measuring and adjusting refrigerant charge in residential and light-commercial air conditioners and heat pumps. The EPA-sponsored report, mentioned in Chapter 1, states that 74% of split-system air conditioners contained a refrigerant charge that was more than 5% higher or lower than the manufacturer's specifications. The report's authors estimate that correcting the charge would save an average of 12% of energy used for cooling.

The type of expansion valve in the air conditioner determines what testing method — superheat or subcooling — to use. Units with fixed-orifice expansion devices require superheat testing. Units with TXVs require subcooling testing. All testing and subsequent addition or removal of refrigerant should be performed by qualified and EPA-licensed refrigeration technicians.

6.3.1 Preparations for Charge Testing

Refrigerant-charge testing and adjustment should be done after airflow measurement and improvement and after duct testing and sealing. The logic behind this sequence is that airflow should be adequate before duct sealing is done in case you have to add or enlarge ducts. Manufacturers recommend that adequate airflow be verified before charge is checked and adjusted. See *"Measuring Airflow" on page 118.*

Clean the condenser coil before testing and adjusting refrigerant charge. After cleaning the coil with outdoor coil cleaner, let the coil dry thoroughly. Otherwise suction and head pressures may be abnormally low.

Required Equipment

For withdrawing refrigerant, use a U.S. Department of Transportation (D.O.T.) recovery cylinder. You'll usually be adding refrigerant from a R-22 or R-410A cylinder, however you can put recovered refrigerant back into the same refrigeration system from the recovery cylinder after evacuating that system.

Required equipment includes a refrigeration gauge set and a digital thermometer. The digital thermometer should have a cloth covering for the tip of one of its thermocouples, which is wetted to measure wet-bulb temperature of air entering the evaporator. A sling psychrometer can also be used to measure wet-bulb temperature of air entering the evaporator.

6.3.2 Evaporator Superheat Test for Refrigerant Charge

Adjusting the refrigerant charge to produce the recommended superheat temperature, based on current indoor and outdoor temperatures optimizes system performance and efficiency. Superheat is an good indicator of correct charging for air conditioners and heat pumps with capillary-tube or fixed-orifice expansion devices, operating in the cooling mode.

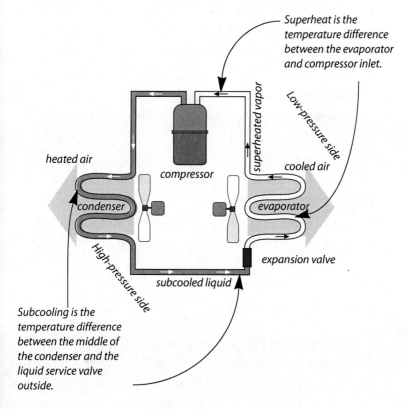

Superheat is the temperature difference between the evaporator and compressor inlet.

Subcooling is the temperature difference between the middle of the condenser and the liquid service valve outside.

Use this test only for fixed orifice or capillary-tube systems and not for thermostatic expansion value (TVX) systems. This test should only be done when the outdoor temperature is less than 60°F.

1. Before checking charge, test and verify adequate airflow, using procedures on *page 102.*

2. Measure the dry bulb temperature of the air entering the outdoor coil.

3. Measure the wet bulb temperature of the return air at the air handler.

4. Determine the recommended superheat temperature from a superheat table.

5. Measure the compressor-suction pressure at the suction-service valve. Add 2 pounds per square inch of gauge pressure (psig) for line losses between the evaporator and compressor. Then convert this adjusted pressure to a boiling-point temperature using temperature-pressure tables.

6. Measure the suction-pipe temperature at the suction service valve by taping and foam-insulating a thermocouple there.

7. Subtract the boiling-point temperature determined in (5) from the measured temperature in (6). This is the actual superheat temperature.

8. If the actual superheat is greater than the recommended superheat obtained from the table (4) by more than 5°F, add 2-4 ounces of refrigerant. Then wait at least ten minutes and repeat this superheat procedure.

9. If the actual superheat is less than the ideal by more than 5°F, remove 2-4 ounces refrigerant, and wait at least ten minutes before measuring superheat again. Refrigerant must be removed into a Department-of-Transportation-approved (DOT-approved) recovery

cylinder, either empty or containing the same refrigerant as the system.

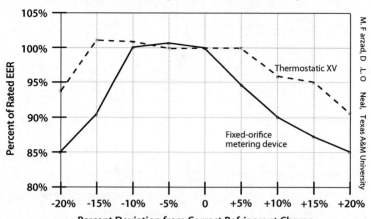

**Comparison of TXV and Fixed-Orifice XV:
EER versus Charge at 95°F Outdoor Temperature**

Energy efficiency ratio (EER) degrades: Fixed-
orifice expansion devices are severely affected by incorrect charge. Thermostatic
expansion valves partially compensate for charge that is too high or low.

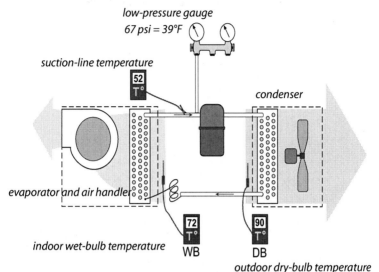

low-pressure gauge
67 psi = 39°F

suction-line temperature
52 T°

condenser

evaporator and air handler

indoor wet-bulb temperature
72 T° WB

90 T° DB

outdoor dry-bulb temperature

Superheat test: Superheat is the heat added to the evaporating vapor to ensure that no liquid enters the compressor. For a fixed-orifice system this value varies with outdoor temperature and indoor temperature and humidity.

Low-pressure gauge reads 67 psi, which corresponds to 39°F evaporating temperature (R-22). Superheat is 52° − 39° = 13°F. Ideal superheat from *"Ideal Superheat Values for Different Indoor and Outdoor Conditions" on page 159* is 24°. This system is overcharged.

Limitations of the Superheat Test

First, superheat won't be accurate unless airflow is around 400 cfm per ton, so airflow should be measured and improved, if inadequate. Sometimes you can't charge by superheat because of either low or high outdoor temperature. Superheat disappears at very high outdoor temperatures, and charge-checking at these temperatures is not recommended. As outdoor temperature rises, system pressures rise, and refrigerant flow rate through the fixed orifice increases until flooding may occur. Therefore, some superheat is desirable, even at high outdoor temperatures to protect the compressor from liquid refrigerant. Providing 1 to 5°F of superheat—even at high outdoor temperatures where superheat values aren't listed—would create a slight undercharge, which would protect the compressor.

Table 6-1: Ideal Superheat Values for Different Indoor and Outdoor Conditions

Measured Return-Air Wet Bulb	Outdoor Condenser Entering Dry Bulb F°										
	65	70	75	80	85	90	95	100	105	110	115
76	41	39	37	35	33	31	29	27	26	25	23
74	38	36	34	31	30	27	25	23	22	20	18
72	36	33	30	28	26	24	22	20	17	15	14
70	33	30	28	25	22	20	18	15	13	11	8
68	30	27	24	21	19	16	14	12	9	6	-
66	27	24	21	18	15	13	10	8	5	-	-
64	24	21	18	15	11	9	6	-	-	-	-
62	21	19	15	12	8	5	-	-	-	-	-
60	19	16	12	8	-	-	-	-	-	-	-
58	16	13	9	5	-	-	-	-	-	-	-
56	13	10	6	-	-	-	-	-	-	-	-
54	10	7	-	-	-	-	-	-	-	-	-
52	6	-	-	-	-	-	-	-	-	-	-
50	-	-	-	-	-	-	-	-	-	-	-

6.3.3 Subcooling Test to Ensure Proper Charge

Follow manufacturer's instructions for the subcooling test, if available. This test is only to be used for thermal expansion valve (TXV) systems when the outdoor temperature is at least 60°F. The air conditioner or heat pump should be running in the cooling mode for 10 minutes prior to the test.

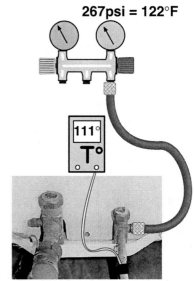

267psi = 122°F

Subcooling: 122°–111°=11°F

1. Measure the liquid pressure at the liquid service valve. Convert this pressure to the condenser saturation temperature, using the correct temperature scale on the gage or temperature-pressure tables for the system's refrigerant.

Subcooling test with R-22: The temperature drop between the condenser and the liquid line is called subcooling and is an test for correct charge in systems with TXVs. The condensing temperature is derived from measured liquid pressure.

2. Measure the temperature of the liquid refrigerant leaving the condenser.

3. Subtract the liquid-refrigerant temperature measured in (2) from the condensing temperature determined in (1). This is the subcooling.

4. Find the correct subcooling from the permanent sticker inside the condenser unit, from manufacturer's literature, or from a manufacturer's slide rule.

5. Add refrigerant if the measured subcooling temperature is 3°F or more below the recommendation. Withdraw refrigerant if the subcooling temperature is 3°F or more greater than recommended. Refrigerant must be

removed into an empty DOT-approved recovery cylinder or one containing the same refrigerant as the system.

6. Allow the system to run for 10 minutes to adjust to the new operating conditions. Repeat the subcooling procedure, adding or removing refrigerant as necessary until the measured subcooling temperature is within 3°F of the manufacturer's recommendation or is between 10° and 15°F.

6.3.4 Weigh-in Test to Ensure Proper Refrigerant Charge

Weighing-in is the preferred method of achieving the correct charge. Weigh in refrigerant whenever you are charging any of the following.

- New installations,

- Systems where the refrigerant has leaked out,

- To correct refrigerant charge if found to be incorrect after checking superheat or subcooling, or

- To remove existing refrigerant in an EPA-approved manner and recharge the system by weighing in the correct amount of refrigerant whenever superheat or subcooling tests can't be employed.

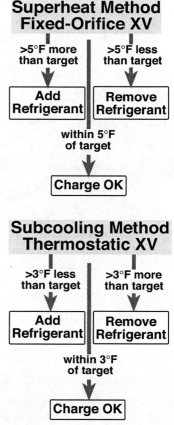

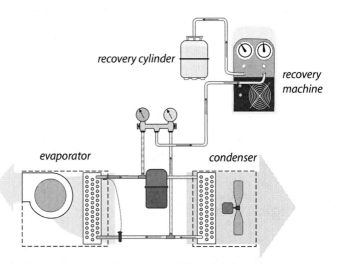

recovery cylinder

recovery machine

evaporator

condenser

Refrigerant recovery. Most of the refrigerant will flow from the system to a recovery cylinder as a liquid, while being filtered by the recovery machine. The recovery machine can recover the most of the remaining refrigerant as a vapor by pulling a vacuum of 10 inches of mercury.

How to Weigh In

Follow these procedures to evacuate the existing charge and weigh in the correct one.

1. Follow the recovery machine's operating instructions for connecting hoses.

2. Remove the refrigerant with the recovery machine. The recovery machine should pull a vacuum of at least 10 inches of mercury. Recover the refrigerant into a DOT-approved cylinder, noting the weight of refrigerant recovered.

3. Using a vacuum pump, evacuate the system to 500 microns to remove moisture and impurities.

4. Determine the correct charge by reading it from the nameplate. Determine the length of lineset that the nameplate charge assumes from manufacturer's literature or by contacting a manufacturer's representative.

5. Accurately measure the unit's installed lineset length. Depending on whether the existing lineset is longer or shorter than the manufacturer's assumed lineset length, add or subtract ounces of refrigerant, based on the manufacturer's specifications of refrigerant ounces per foot of suction and liquid line.

6. Connect the EPA-approved recovery cylinder or disposable cylinder to the gauge manifold. Place the cylinder on an electronic scale and zero the scale. During charging, be very careful not to bump or otherwise disturb the scale or cylinder. The scale could reset itself, forcing you to evacuate and start all over.

7. To prepare for liquid charging, connect the common port of the gauge manifold to the liquid valve of the recovery cylinder. If using a disposable cylinder, turn the cylinder upside down after connecting the common port to the cylinder valve.

8. With the compressor off, open the cylinder's valve and suction service valve, and let the liquid refrigerant flow in.

Purging Refrigerant for a Gauge Hose

When you connect your high-pressure gauge to the high-pressure liquid service valve, the hose fills with liquid refrigerant. Rather than releasing this refrigerant into the atmosphere or carrying it around to contaminate another system, run it back into the low-pressure suction service valve. Modern hoses hold refrigerant under pressure when they are disconnected, so disconnect the high-pressure hose. Open the common chamber of the gauge set to the suction service valve. Then open the common chamber to the high-pressure hose, slowly letting the liquid refrigerant hiss through the chamber, vaporizing before reaching the low-pressure service valve or nearby compressor.

9. If liquid stops flowing before the correct charge has entered, reconfigure the gauge manifold and cylinder to charge with vapor through the suction service valve.

10. With the compressor running, add the remaining refrigerant as a vapor. Before opening a path between the cylinder and the system, check the low-pressure gauge to make sure the cylinder pressure is higher than the system's suction pressure.

11. Weigh in the remainder of the charge carefully and slowly to avoid slugging the compressor.

12. Check performance after 10 minutes of operation using superheat test or subcooling test.

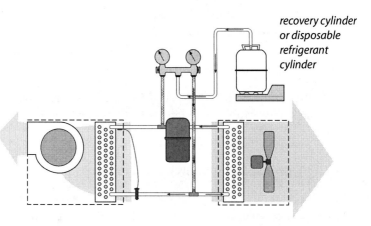

recovery cylinder or disposable refrigerant cylinder

Liquid charging: Most of the captured refrigerant should flow into the evacuated system as a liquid through the liquid service valve.

Limitations of the Weigh-In Method

Ideally, the service technician has the manufacturer's literature, which specifies the lineset length, assumed by the manufacturer, and the amount of refrigerant required for each foot of suction and liquid line. You may also need to know the weight of refrigerant contained in the indoor coil. The weigh-in method can't be performed accurately without this information or without accurately measuring the lineset length. Sometimes, the difficulty of obtaining the manufacturer's specifications makes the weigh-in method a poor choice because the difference between the manufacturer's choices of linesets (15, 20, and 25 feet) can result in a charge no more accurate than ±5 to 15 ounces, which is unacceptable.

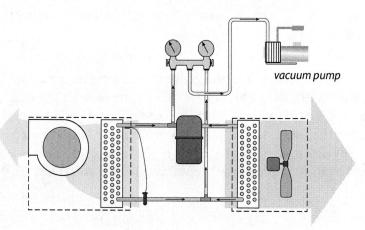

vacuum pump

Vacuum-pump evacuation: The last remains of the system's refrigerant are vented into the atmosphere along with air and moisture, leaving the system clean and ready for charging.

6.4 HEAT PUMP ENERGY EFFICIENCY

Heat pumps move heat with refrigeration rather than converting it from the chemical energy of a fuel. Like air conditioners, air-source heat pumps are available as centralized units with ducts or as room units. Heat pumps are 1.5 to 3 times more

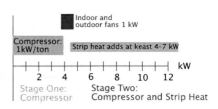

Is the auxiliary heat activated? Using an ammeter and the nameplate data on the heat pump, a technician can know when and if the auxiliary heat is activated.

efficient than electric furnaces. Heat pumps can provide competitive comfort and value with combustion furnaces but they must be ducted and installed with exceptional care and planning.

An air-source heat pump is almost identical to an air conditioner, except for a reversing valve that allows refrigerant to follow two different paths, one for heating and one for cooling. Heat pumps are also equipped with auxiliary electric resistance heat, called strip heat. The energy efficiency of a heat pump is largely determined by how much of the heating load can be handled by the compressor without the aid of the strip heat.

Testing central heat pumps during the summer follows the same procedures as testing central air conditioners and described on *page 154.* Testing heat pumps in the winter is more difficult and some specifications follow.

✓ Look for a temperature rise of around half the outdoor temperature in degrees Fahrenheit.

✓ Check for auxiliary-heat operation with an ammeter, using the chart shown here. Heat pumps should have two-stage thermostats designed for heat pumps. The first stage is compressor heating and the second stage brings on auxiliary heat to aid the compressor.

✓External static pressure should be 0.5 IWC (125 pascals) or less for older, fixed-speed blowers and less than 0.8 IWC (200 pascals) for variable-speed and two-speed blowers. Lower external static pressure is better. Take necessary steps to reduce external static pressure, such as enlarging branch ducts, installing additional supply and return ducts, and following other specifications on *page 124.*

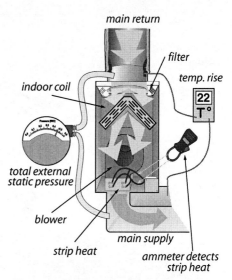

main return

filter

temp. rise

indoor coil

total external static pressure

blower

strip heat

main supply

ammeter detects strip heat

Heat pump: The air handler contains a blower, indoor coil, strip heat, and often a filter. Static pressure and temperature rise are two indicators of performance.

✓Supply ducts should be sealed and insulated after the airflow has been verified as adequate. Return ducts should be sealed too.

Most residential central heat pumps are split systems with evaporator and air handler indoors and condenser and compressor outdoors. The illustrations show features of an energy-efficient heat pump installation.

Individual room heat pumps are more efficient since they have no ducts and are factory-charged with refrigerant. Room heat pumps are discussed with room air conditioners on *page 153.*

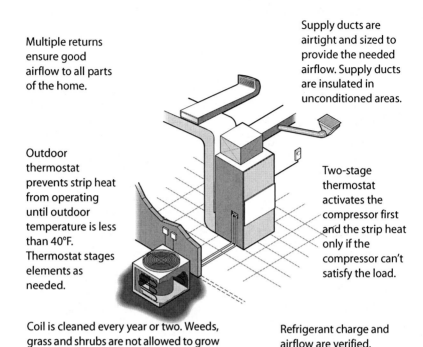

Multiple returns ensure good airflow to all parts of the home.

Supply ducts are airtight and sized to provide the needed airflow. Supply ducts are insulated in unconditioned areas.

Outdoor thermostat prevents strip heat from operating until outdoor temperature is less than 40°F. Thermostat stages elements as needed.

Two-stage thermostat activates the compressor first and the strip heat only if the compressor can't satisfy the load.

Coil is cleaned every year or two. Weeds, grass and shrubs are not allowed to grow within 3 feet on all sides.

Refrigerant charge and airflow are verified.

6.5 EVAPORATIVE COOLERS

Evaporative coolers (also called swamp coolers) are energy efficient cooling strategy in dry climates. An evaporative cooler is a blower and wetted pads installed in a compact louvered air handler.

Evaporative coolers employ different principles from air conditioners because they reduce air temperature without actually removing heat from the air. They work well only in climates where the summertime relative humidity remains less than 50%. They compare in performance to an air conditioner with a SEER between 30 and 40, which is 2 to 3 times the SEER of the most efficient air conditioners.

An evaporative cooler can be mounted on a roof, through a window or wall, or on the ground. The cooler can discharge air directly into a room or hall or it can be connected to ducts for distribution to numerous rooms.

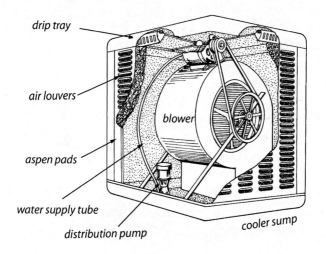

6.5.1 Evaporative Cooler Operation

The evaporative cooler's blower moves outdoor air through water-saturated pads, reducing the air's temperature to below the indoor air temperature. This evaporatively cooled outdoor air is blown into the house by the blower, pushing warmer indoor air out through open windows or dedicated up-ducts.

A water pump in the reservoir circulates water through tubes into a drip trough, which then drips water into the thick pads. A float valve connected to the home's water supply keeps the reservoir supplied with fresh water to replace the water evaporated.

Opening windows in occupied rooms, and closing windows in unoccupied rooms concentrates the cooling effect where residents need it. Open the windows or vents on the leeward side of the home to provide approximately 1 to 2 square feet of opening for each 1,000 cfm of cooling capacity. Experiment to find the right windows to open and how wide to open them. If the windows are open too wide hot air will enter. If the windows are not open far enough humidity will rise, and the air will feel sticky.

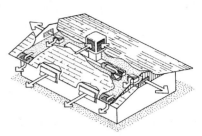

Air circulation: Outdoor air is sucked through the evaporative cooler and blown into the home pushing house air out the partially open windows.

6.5.2 Evaporative Cooler Sizing and Selection

Evaporative coolers are rated in cubic feet per minute (cfm) of airflow they deliver. Airflow capacity ranges from 2000 to 7000 cfm. Recommendations vary from 2-to-3 cfm per square foot of floor space for warm dry climates and 3-to-4 cfm/sf for hot desert climates.

6.5.3 Evaporative Cooler Installation

Evaporative coolers are installed in two ways: the cooler outlet blows air into a central location, or the cooler outlet joins ductwork which distributes the cool air to different rooms in the house. Single outlet installations work well for compact homes. Ducted systems work better for more spread-out homes.

Evaporative coolers are installed on roofs, on concrete slabs, or on platforms. These coolers vent their cool air through windows or ducts cut through walls.

The best place for a horizontal-flow evaporative cooler is in the shade on the windward side of the home. Consider the following installation specifications.

✓ The cooler should have a fused disconnect and water shut-off nearby.

✓ The cooler should have at least 3 feet of clearance all around them for airflow and maintenance access.

✓ Using thermostats to control evaporative coolers minimizes energy use, water use, and maintenance. Thermostats also reduce the chance of over-cooling with unnecessary night-time operation. Using a 24-volt transformer and thermostat is better than using a line-voltage thermostat, which allows more temperature variation.

✓ Evaporative coolers often have two speeds for cooling or venting. The vent settings activate the blower but not the pump, to use the cooler as a whole-house fan at night and during mild weather.

✓ Selecting a control, equipped with a pump-only setting, permits flushing dirt out of the pads before activating the blower after the cooler has been off for days or weeks.

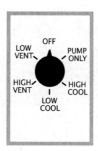

Evaporative cooler control: Controls like this provide comfort in a variety of summer weather conditions.

✓ Choose a cooler that has a bleed tube or sump pump to drain dirty water from the sump.

✓ Installing the cooler with weighted dampers allows easier changeover from evaporative cooling and either heating or air conditioning.

Evaporative coolers produce high air flows; the ductwork connected to them should be sized appropriately to the cooler's airflow rating. The cooler's supply outlet can supply one or more registers through a dedicated duct system. Or, the supply outlet can connect to ducts that join to furnace or air conditioner ducts.

Coolers sharing ducts with forced-air furnaces require dampers to prevent heated furnace air from blowing into the idle evaporative cooler during the winter, and also to prevent moist cool air from blowing into the furnace during the summer. Moist cool air can condense and cause rust inside the furnace. These shared systems must be installed with great care. The dampers often stick in an open position even with careful installation.

Up-ducts are ceiling vents that exhaust warm air as the evaporative cooler pushes cooler air in. Up-ducts are preferred by home owners who don't like leaving windows open for security or privacy reasons. Up-ducts also help maintain a positive pressure in the home, preventing wind-driven hot air from entering through open windows. It is essential to have adequate attic ventilation when using up-ducts. Attic vents should have 1 to 1.5 times the net free area of the up-ducts.

6.5.4 Evaporative Cooler Maintenance

Evaporative coolers see a lot of water, air, and dirt during operation. Dirt is the enemy of evaporative cooler operation. Evaporative coolers process a lot of dirt because their aspen pads are good filters for dust-bearing outdoor air. Evaporative coolers may cool better and filter better when the aspen pads are doubled up.

Airborne dirt that sticks to the cooler pads washes into the reservoir. Most evaporative coolers have a bleed tube or a separate pump that changes the reservoir water during cooler operation to drain away dirty water. A cooler may still need regular cleaning, depending on how long the cooler runs and how well the dirt-draining system is working. Be sure to disconnect the electricity to the unit before servicing or cleaning it.

Older steel cooler sumps are lined with an asphaltic paint and flexible asphaltic liners, but the newer ones have factory powder coatings that are far superior and less environmentally harmful. Don't paint or install asphalt liners in a powder-painted cooler sump because asphaltic material won't stick to the factory finish. Newer evaporative coolers have plastic housings and plastic sumps that don't need a coating.

Observe these general specifications for maintaining evaporative coolers.

✓ Aspen pads can be soaked in soapy water to remove dirt and then rotated to distribute the wear, dirt, and scale, which remains entrained after cleaning. Clean louvers in the cooler cabinet when you clean or change pads. Replace the pads when they become unabsorbent, thin, or loaded with scale and entrained dirt.

✓ If there is a bleed tube, check discharge rate by collecting water in a cup or beverage can. You should collect a cup in three minutes or a can in five minutes.

✓ If the cooler has two pumps, one is a sump pump. It should activate to drain the sump every five to ten minutes of cooler operation.

✓ If there is any significant amount of dirt on the blower's blades, clean the blower thoroughly. Clean the holes in the drip trough that distributes the water to the pads.

✓ The reservoir should be thoroughly cleaned each year to remove dirt, scale, and biological matter. You can gather silt and debris using two old hand towels or rags working together from the corners of the sump pushing the dirt and silt into the sump drain or into a bucket.

✓ Pay particular attention to the intake area of the circulating pump during cleaning. Debris can get caught in the pump impeller and stop the pump.

✓Check the float assembly for positive shutoff of water when the sump reaches its level. Repair leaks and replace a leaky float valve.

✓Investigate signs of water leakage and repair water leaks.

Outdoor Relative Humidity %

Outdoor Temperature F	2	5	10	15	20	25	30	35	40	45	50	55	60	65	70
75	54	55	57	58	59	61	62	63	64	65	66	67	68	69	70
80	57	58	60	62	63	64	66	67	68	71	72	73	74	76	76
85	61	62	63	65	67	68	70	71	72	73	74	75	76	77	79
90	64	64	67	69	70	72	74	76	77	78	79	81	82	83	84
95	67	68	70	72	74	76	78	79	81	82	84	85	87		
100	69	71	73	76	78	80	82	83	85	87	88				
105	72	74	77	79	81	84	86	88	89						
110	75	77	80	83	85	87	90	92							
115	78	80	83	86	89	91	94								
120	81	83	86	90	93	95									
125	83	86	90	93	96										

An evaporative cooler with good pads and adequate airflow should give the temperatures listed here, depending on outdoor temperature and relative humidity.

Chapter 7: Mechanical Ventilation

Properly designed ventilation systems perform several tasks.

- They protect human health by diluting airborne pollutants and by supplying fresh air.

- They improve comfort by eliminating odors and reducing drafts.

- And they preserve structures by controlling airborne water vapor, thus preventing condensation.

Mechanical ventilation systems are different than the passive vents often installed in attics and crawl spaces to control moisture and heat. Mechanical ventilation systems move air through the home using fans and ducts. Electric controls regulate ventilation airflow according to need.

The home's airtightness level determines its need for mechanical ventilation. Many homes have been built over the years that are ventilated only by simple bathroom and kitchen fans, open-able windows, and air leakage through the building shell. This uncontrolled ventilation is sometimes sufficient remove moisture and air pollutants, and sometimes not. Mechanical ventilation becomes more of a necessity as homes become more airtight.

Types of Ventilation

Three types of mechanical ventilation are defined in this chapter.

1. Spot ventilation for removing moisture and odors at their source.

2. Central ventilation for moderately airtight homes uses exhaust fans or outdoor-air ducts connected to the air handler's return plenum. With these options, air leaks provide make-up air and pressure relief to balance the intake with exhaust.

3. Balanced central ventilation for very airtight homes employs two fans, one for intake and one for exhaust. These systems usually contain a heat exchanger to recover energy from the exhaust air.

7.1 SPOT VENTILATION

All kitchens and bathrooms should have exhaust fans. All exhaust fans should be installed as follows:

✓ Duct range hoods should always be ducted to the outdoors. Don't install re-circulating range hoods and consider replacing them with ducted hoods. The re-circulating hood filters can't remove pollutants such as carbon dioxide, carbon monoxide, or water vapor.

✓ Install high-quality exhaust fans with a low noise rating.

The success of spot ventilation and whole-house ventilation depends on how much noise the fans make. Occupants may not use the fans or may disconnect automatic fan controls if the fans are too noisy. The sound output of exhaust fans is rated in sones, and these ratings vary from about 5 sones for the noisiest residential exhaust fans to about 0.5 sones for the quietest fans. An Energy Star-labelled exhaust fan ensures a quiet, energy-efficient exhaust fan.

7.2 CHOOSING WHOLE-HOUSE VENTILATION SYSTEMS

Ventilation systems must be matched to the home. A home may require only simple exhaust fans in a bathroom, kitchen, or common area. Very tight homes may require a balanced central ventilation system.

The first step in considering a whole-house ventilation system is to evaluate the home's current ventilation level.

Table 7-1: Choosing Whole-House Ventilation Systems

Type of System		House Pressure Influence	Forced-Air System Required	Heat Recovery
Exhaust Ventilation	Spot Ventilators	Negative	No	No
	Central Ventilator	Negative	No	No
Furnace Supply Ventilation	Furnace Supply	Positive	Yes	No
Central Balanced Ventilator	Simplified HRV	Balanced	Yes	Yes
	Ducted-Exhaust HRV	Balanced	Yes	Yes
	Fully Ducted HRV	Balanced	No	Yes

7.3 EVALUATING HOME VENTILATION LEVELS

Most North American homes use only air leakage for ventilation, a practice which is now considered inferior to installing a whole-house ventilation system. The American Society of Heating, Refrigeration, and Air Conditioning Engineers (ASHRAE) sets ventilation standards. There are two ASHRAE standards currently being used in North America. The first is ASHRAE 62-1989 and the second is ASHRAE 62.2-2003. Both ASHRAE standards are covered in this section.

ASHRAE 62-1989 assumes that air leakage is a legitimate way to provide whole-house ventilation. This ASHRAE standard establishes a minimum lower limit for air leakage, which unfortunately is known by at least five different terms.

- Building tightness limit (BTL)
- Building airflow standard (BAS)

- Minimum ventilation level (MVL)
- Minimum ventilation guideline (MVG)
- Minimum ventilation requirement (MVR)

Under ASHRAE 62-1989, air leakage is measured by a blower door in CFM at 50 pascals (CFM_{50}).

The newer standard, ASHRAE 62.2-2007, requires fan-powered ventilation in all but very leaky homes and homes in very mild climates. Homes throughout North America probably need to become more airtight to cope with rising energy costs. According to ASHRAE 62.2-2007 these more airtight homes require mechanical whole-house ventilation systems. ASHRAE 62,2-2007 calculates the minimum fan-powered airflow that could be referred to by any of the last three terms on the above list.

The 2006 International Energy Conservation Code (IECC) requires the ASHRAE 62.2-2003 procedure for sizing whole-house ventilation systems and the ASHRAE 62-1989 procedure ($0.35 ACH_n$) for verifying an adequate natural ventilation rate for homes without whole-house ventilation systems.

7.3.1 ASHRAE 62-1989: Building Tightness Limits (BTL)

Air leakage must provide fresh outdoor air when no mechanical ventilation system exists. Without fan-powered ventilation, air leaks are the home's only source of fresh air to dilute pollutants. The air leakage is measured with a blower door. Follow these steps to determine the building tightness limit (BTL), which is also called four other names. The following is the ASHRAE 62-1989 procedure for determining the BTL.

1. Obtain the number of occupants by each of the following ways, and then use the largest number: a) actual number of occupants, and b) number of bedrooms plus one.

2. Find the zone from the map shown here.

3. Decide whether the building is well-shielded from wind, directly exposed to wind, or "normal," which means somewhere in-between exposed and well-shielded.

4. Find the factor "n" where the column representing the building's number of stories meets the row representing your location and the building's shielding. (This factor converts 50-pascal airflow to natural airflow and vice versa.)

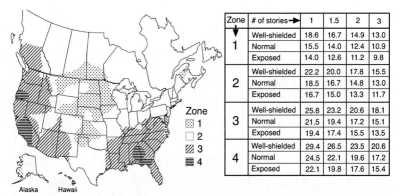

Zone	# of stories ➡	1	1.5	2	3
1	Well-shielded	18.6	16.7	14.9	13.0
	Normal	15.5	14.0	12.4	10.9
	Exposed	14.0	12.6	11.2	9.8
2	Well-shielded	22.2	20.0	17.8	15.5
	Normal	18.5	16.7	14.8	13.0
	Exposed	16.7	15.0	13.3	11.7
3	Well-shielded	25.8	23.2	20.6	18.1
	Normal	21.5	19.4	17.2	15.1
	Exposed	19.4	17.4	15.5	13.5
4	Well-shielded	29.4	26.5	23.5	20.6
	Normal	24.5	22.1	19.6	17.2
	Exposed	22.1	19.8	17.6	15.4

Zone
▨ 1
☐ 2
▨ 3
▤ 4

Alaska   Hawaii

Finding the n-value: Find your zone from the map. Pick the correct column by the number of stories in the building. Then decide how exposed the building is and find the n-value.

5. Calculate the BTL using the following formulas. Calculate the BTL both ways and use the largest CFM_{50} number.

$$\textbf{BTL } CFM_{50} = 15 \text{ cfm} \times \text{\# occupants} \times n$$

$$\textbf{BTL } CFM_{50} = \frac{0.35 \text{ ACH50} \times \text{volume} \times n}{60}$$

Pollution control and whole-house fan-powered ventilation are priorities for homes with tested blower door measurements below the BTL. The importance of pollution control and whole-house ventilation become more urgent and important with yes answers to the following questions.

- Are sources of moisture like ground water, humidifiers, water leaks, or unvented space heaters causing indoor dampness, high relative humidity, or moisture damage?

- Do occupants complain or show symptoms of building-related illnesses?

- Are there combustion appliances in the living space?

- Are the occupants smokers?

7.3.2 ASHRAE 62.2-2007 Ventilation Standard

To comply with ASHRAE 62.2-2007, you can use either the formula or the table to determine the minimum ventilation level (MVL) in CFM of fan-powered airflow. ASHRAE 62.2-2007 doesn't consider blower door measurements or n-values to compute the MVL, which is provided by a whole-house ventilation fan or fans. Use the method outlined here to size whole-house ventilation systems.

Obtain the number of occupants for this formula by the actual number of occupants or the number of bedrooms plus one, whichever number is higher.

MVL (CFM) = (7.5 cfm x # occupants) + (0.01 x floor area)

The table below displays the CFM of required continuous fan flow, depending on the number of bedrooms and the conditioned floor area of the home.

Table 7-2: Fan Sizes for Homes with Average Air Leakage

Floor Area (ft^2)	No. of Bedrooms				
	0-1	2–3	4–5	6–7	>7
< 1500	30	45	60	75	90
1501–3000	45	60	75	90	105
3001–4500	60	75	90	105	120
4501–6000	75	90	105	120	135
6001–7500	90	105	120	135	150
> 7500	105	120	135	150	165
From ASHRAE Standard 62.2-2007					

7.4 INSTALLING WHOLE-HOUSE VENTILATION

This section discusses three options for design of whole-house ventilation systems.

1. Exhaust ventilation.

2. Supply ventilation.

3. Balanced ventilation.

7.4.1 Exhaust Ventilation

Exhaust ventilation systems use an exhaust fan to remove indoor air, which is replaced by infiltrating outdoor air. Better air distribution is achieved by using a remote fan that exhausts air from several rooms through small (3-to-4 inch) diameter ducts.

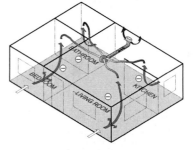

Multi-port exhaust ventilation: A multi-port ventilator creates better fresh-air distribution than a single central exhaust fan.

In smaller homes, it is adequate to install a high-quality ceiling exhaust fan, preferably in a central location. For simplicity, the central ventilation fan should run continuously, and local exhaust fans should be employed as usual, to remove moisture and odors as needed. Continuous ventilation allows for a small fan size that will minimize the depressurization compared to intermittent ventilation with a larger fan.

To replace a bathroom exhaust fan with an exhaust ventilation fan, the new fan should run continuously on low speed for whole-house ventilation. This fan should also have a high speed that occupants can use to remove moisture and odors from the bathroom or kitchen quickly.

Exhaust ventilation systems are inexpensive and easy to install, but it isn't possible to recover heating and cooling energy or to control the source of incoming air. Since most exhaust ventilation systems don't have filters, dust collects in the fan and ducts and must be cleaned out every year or so to preserve the design airflow. Exhaust ventilation systems create negative pressure so they aren't appropriate for homes with fireplaces or other open-combustion appliances.

Exhaust systems create negative pressure within the home, drawing air in through leaks in the shell. This keeps moist indoor air from traveling into building cavities, reducing the likelihood of moisture accumulation in cold climates during the winter months. In hot and humid climates, however, this depressurization can draw outdoor moisture into the home.

Fan Specifications

Continuous ventilation is highly recommended because it simplifies ventilation design and control and also minimizes depressurization by allowing selection of the minimum-sized fan. Exhaust fans, installed as part of weatherization or home-performance work, must vent to outdoors and include the following.

Specifying exhaust fans: Specify quiet energy-efficient fans.

✓The ENERGY STAR® seal.

✓A weatherproof termination fitting.

✓A backdraft damper, installed in the fan housing or termination fitting.

✓Noise rating and ventilation efficacy as specified here.

Table 7-3: Fan Capacity, Maximum Noise Rating, & Efficacy

Fan Capacity	Noise Rating (sones)	Efficacy cfm/Watt
<50 CFM	≤1 sone	≥2.8
50–100 CFM	≤1.5 sones	≥2.8
>100 CFM	≤2.0 sones	≥2.8

7.4.2 Supply Ventilation

Supply ventilation uses your furnace or heat pump as a ventilator. A 5-to-10 inch diameter duct is connected from outdoors to the air handler's main return duct. This outdoor-air supply duct may have a motorized damper that opens when the air-handler blower operates. This outdoor air is then heated or cooled in the air handler before its delivery into the living space.

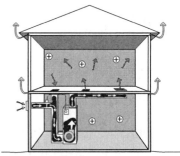

Supply ventilation: A furnace or heat pump is used for ventilation with a control that ensures sufficient ventilation.

At least one manufacturer makes a control for running the furnace blower and damper for ventilation. The control activates both the damper and blower as necessary to provide sufficient ventilation. To accomplish this, the control logs the time that the blower operated for heating or cooling. It calculates whether this time is sufficient to bring enough intake air for ventilation. If the intake air isn't sufficient, the control operates the blower without heating or cooling to bring in additional intake air.

Supply ventilation through the air handler doesn't operate continuously as with exhaust ventilation because the air handler's blower is too large and would over-ventilate the home. Supply ventilation isn't appropriate for very cold climates because it pressurizes the home, pushing indoor air through exterior walls where moisture can condense on cold surfaces.

7.4.3 Balanced Ventilation Systems

Balanced ventilation systems provide fresh air via planned pathways and do the best job of controlling pollutants in the home compared to supply ventilation and return ventilation.

Designed balanced systems move equal amounts of air into and out of the home. Most balanced systems incorporate heat recovery ventilators that reclaim some of the heat and/or moisture from the exhaust air stream.

Balanced systems, when operating properly, reduce many of the safety problems and moisture-induced building damage that is possible with unbalanced ventilation. Balanced systems are not trouble-free, however. Proper design, installation, and maintenance are required for effective operation. Testing and commissioning is vital during both the initial installation and periodic service calls.

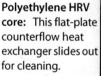
Polyethylene HRV core: This flat-plate counterflow heat exchanger slides out for cleaning.

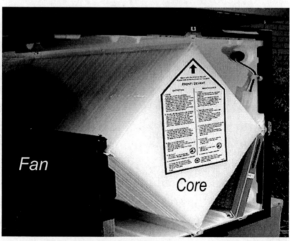

Heat recovery ventilators (HRVs) or energy recovery ventilators (ERVs) are the most common balanced whole-house ventilation systems.

HRVs or ERVs can offset some of the energy loss from exhausted air and impart that energy to incoming air. The savings from HRVs or ERVs over ventilators without heat/energy recovery are greatest where outdoor temperatures are most severe. The HRV or ERV heat exchanger is usually a heat exchanger, inside which the supply and exhaust airstreams pass one another and exchange heat.

Balanced ventilation systems can be designed and installed in three common variations.

1. Fully ducted systems

2. Ducted-exhaust balanced systems

3. Simplified balanced systems

Variation 1: Fully Ducted Balanced Systems

The most effective central ventilation systems employ dedicated ductwork for both supply and exhaust air. Fully ducted systems are installed independently of other forced-air ducting. All the ducting leads to a central ventilator that includes an HRV or ERV to reclaim heat.

This gives the designer a high level of control over airflow and house pressure. They are most easily installed in new construction and are more difficult to install during weatherization or renovation.

High quality ductwork is a critical component of successful ventilation systems. Ducts should be sized large enough to minimize static pressure and reduce noise, and hard metal ducting used wherever possible. All seams and joints should be sealed using mastic or metallic tapes. Exhaust grilles should be installed near the sources of contaminants in bathrooms, kitchens, or other areas where other pollutant-producing activities take place.

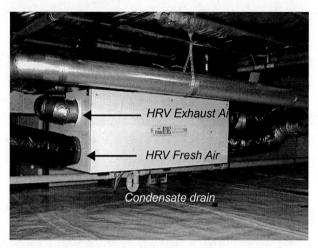

HRV Exhaust Air

vanEE

HRV Fresh Air

Condensate drain

Fully-ducted heat recovery ventilator: Matched exhaust and supply fans provide balanced airflow. Dedicated exhaust ducting collects pollutants from bathrooms and kitchen. Supply ducting carries fresh air to bedrooms and central living areas. A heat-recovery core reduces energy loss from exhausted air.

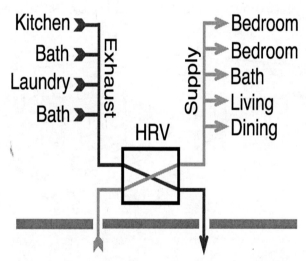

Kitchen ⟩
Bath ⟩
Laundry ⟩
Bath ⟩

Exhaust

Supply

HRV

Bedroom
Bedroom
Bath
Living
Dining

Fully-ducted central ventilator: Fully ducted systems do the best job of collecting pollutants. They are installed independently of heating and cooling systems, and work well in homes with hydronic or electric baseboard heat where no ducting is installed.

Variation 2: Ducted-Exhaust Balanced Systems

Ducted-exhaust systems are connected to central forced-air systems. Dedicated ducting collects pollutants from bathrooms and kitchens. The exhaust air passes through a central HRV before being exhausted to the outdoors. Fresh air is brought in through the HRV, and is introduced to the forced-air system at either the supply or return plenum. Always follow the manufacturer's requirements for mixed air temperature, the location of the fresh air insert, and minimum return air temperature.

The airflow should be balanced in ducted-exhaust systems so house pressures remain close to neutral. Pressure balancing is harder to achieve in ducted-exhaust systems than in fully-ducted systems because of the influence of the forced-air blower. With typical airflows of 50-200 CFM, central ventilators are easily overwhelmed by the 500-1500 CFM airflows of forced-air systems. Design, installation, and commissioning of ducted-exhaust ventilation systems should achieve correct airflows and balanced house pressures.

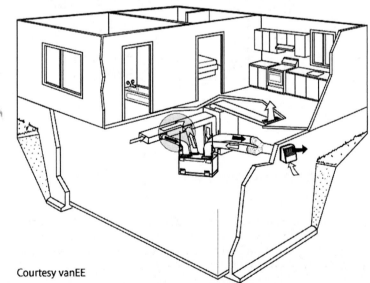

Courtesy vanEE

Ducted exhaust central ventilator: The exhaust is ducted from the pollution-producing rooms into the ventilator. The fresh air comes through the ventilator and into the main return-air duct of the furnace.

Variation 3: Simplified Balanced Systems

Simplified, or volume-ventilation, systems are connected to central forced-air heating or cooling systems. A simplified system is the least-desirable ducting option of the three presented here.

Simplified systems draw exhaust air from the forced-air return air plenum. This exhaust air passes through the central ventilator that includes an HRV or ERV. Most of the exhaust airstream's heat is transferred to supply airstream, and fresh air is re-introduced to the forced-air return ducting. Always follow the manufacturer's recommendation for interlocking the air handler with the fresh air inlet. Simplified systems require the active involvement of the customer because the system fails if maintenance is neglected.

7.5 VENTILATION CONTROL STRATEGIES

Controls let the installer and customers choose when the system runs and how much air it moves. Controls also provide an opportunity to adjust the system performance over time. Installers should review the control scheme during service visits, to assure that the system provides sufficient fresh air for occupants and acceptable moisture control for the building.

Locate the controls in a representative location on a main floor interior wall, and about 60 inches above the floor. Don't install them on an exterior wall, in a drafty location, or in direct sunlight.

7.5.1 Manual Control

Simple on/off manual controls allow occupants to ventilate as needed. These are often used for exhaust fans in bathrooms and kitchens. Their effectiveness relies on the user's perception of air quality.

Manual controls sometimes include count-down or time-delay timers that are activated by occupants and run for a specific period of time. In non-owner occupied homes or other situations where occupant understanding and cooperation is unlikely, fan-delay timers can be run in conjunction with bathroom lights to give a set period of ventilation whenever the bathroom lights are used.

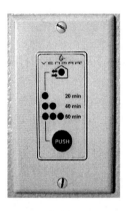

Manual override control: A central heat recovery ventilator, normally operating at low speed, is boosted to high speed by this push-button countdown timer.

7.5.2 Humidity Control

Dehumidistats operate ventilation fans based upon indoor humidity levels. Dehumidistats are used with either simple exhaust fans or central ventilation equipment.

Dehumidistats can be set for a range of humidity levels, and have the advantage of automatic operation that doesn't require much occupant management. Dehumidistats should be set to keep indoor humidity low enough to prevent indoor condensation during the winter. Low-enough humidity varies from 30–50% relative humidity, depending upon the outdoor temperature, effectiveness of windows and insulation, and other factors.

7.5.3 Combination Controls

Central ventilation systems are often operated by a combination of manual and automatic controls. The most common strategy utilizes a multi-speed fan that runs on low or medium speed to provide continuous ventilation. Override switches in the kitchen and bathrooms activate high-speed operation to provide inter-mittent high-speed operation during polluting activities such as cooking, bathing, or cleaning. The total airflow requirement specified by ventilation standards refers to this high-speed oper-ation.

Timers allow the low-speed operation to be set for variable intervals such as 20 minutes on/40 minutes off per hour, 30 on/30 off, or whatever ventilation time is needed. This adjustable interval provides an effective method of matching the ventila-tion capacity to the occupants' needs.

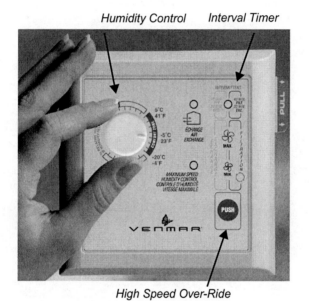

Humidity Control   Interval Timer

High Speed Over-Ride

Central combination control: The system can be controlled by humidity, time interval, or manually.

CHAPTER 8: WATER-HEATING ENERGY SAVINGS

8.1 WATER-HEATER REPLACEMENT

Replacing water heating equipment presents numerous choices especially for managers of multifamily buildings. Direct-fired storage water heaters are the most common water heater in single-family homes. Single-family and multifamily buildings with boilers often have indirect water heaters, which are powered by the same boiler used for space heat. Tankless water heaters and solar water heating are also viable options for both single-family and multifamily buildings.

8.1.1 Determinants of Water-Heating Efficiency

There are two types of efficiency that rate water heaters. The first is the steady-state efficiency that a combustion water heater heats water. The second is energy factor which is a decimal, less than one, representing the energy remaining in water coming out of the water heater divided by the energy input used to heat the water. The energy factor is always less than the steady-state efficiency because energy factor accounts for storage losses. The water heating system efficiency is less than the energy factor because it accounts for distribution losses.

All the different water-heating systems have system- efficiency factors in common, including the following.

- System efficiency increases with decreases in stored water temperature to a minimum of 120°F.

- System efficiency decreases with increases in the distance of fixtures from the storage tank.

- System efficiency increases with increased insulation on hot-water storage tanks.

- System efficiency is directly related to the heat-source's steady-state efficiency.

- Continuous circulation is less efficient than timed circulation, which is less efficient than demand circulation.

8.2 DIRECT-FIRED STORAGE WATER HEATERS

Direct-fired storage water heaters are the most common water heaters found in single-family homes and are also common in multifamily buildings.

8.2.1 Disadvantages of Gas Storage Water heaters

Standard gas storage water heaters have a number of disadvantages compared to advanced storage water heaters, indirect fired water heaters, instantaneous water heaters, and solar water heaters. We mention these disadvantages because of the current market dominance of these standard storage water heaters and the increasing availability of better choices.

Advanced gas water heater: High-efficiency condensing water heaters are expensive but worth the cost to multifamily buildings.

- Standard gas storage water heaters are little more efficient than their predecessors of ten or twenty years ago.

- Gas storage water heaters are known for having weak draft, which leads to spillage and backdrafting. Induced-draft and sealed-combustion models offer safer but not significantly more efficient alternatives.

- Gas and oil storage water heaters lose heat through vertical flue pipes centered inside their upright tanks, while the heaters aren't operating. This creates a large standby heat loss resulting in energy factors of typically less than 0.60.

- Mineral deposits and dirt collects at the bottom of the tank where the heat is applied. These deposits insulate the tank from the heat applied to heat the water.

- The standard glass-lined tanks may not last 5 years with some water chemistry, high water temperatures, and high hot-water use.

8.2.2 Specifications for Gas Storage Water Heaters

Before purchasing a gas storage water heater, consider other alternatives especially in buildings heated by boilers. Indirect water heaters, powered by the building's heating boiler have advantages over direct-fired water heaters.

Several 90+ energy-factor condensing storage water heaters are now available that make sense for multifamily buildings or single family homes with large water-heating demand.

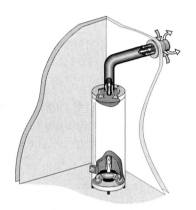

Sealed-combustion water heaters: Safer than conventional gas water heaters, they bring air from outdoors for combustion. Exhaust gases exit through a closed chimney flue.

- In tight homes or homes where the mechanical room is located in a conditioned area, replacement gas or oil water heaters should be either power-draft or sealed-combustion.

- Sealed-combustion water heaters are preferred in tight homes with the water heater installed in a living space.

- Any replacement gas or oil storage water heater for a single-family home should have an energy factor of at least 0.61 or have a minimum of 2 inches of foam insulation around the tank.

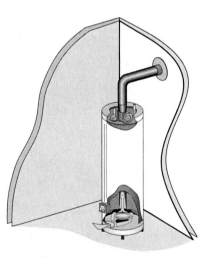

Induced-draft water heaters: A small fan to pull combustion products through the burner and push them into a horizontal vent. Room air is still used to support the combustion process.

- Gas storage water heaters for multifamily buildings should have an energy factor of at least 0.90.

- Replacement water heaters should be wrapped with external insulating blankets for additional savings, unless the manufacturer recommends against installing an external blanket.

8.3 STORAGE WATER HEATING IMPROVEMENT

Gas-, propane-, and oil-fired water heaters must be tested, maintained, repaired, adjusted, and replaced as described in this chapter. Observe the following general specifications concerning water heaters.

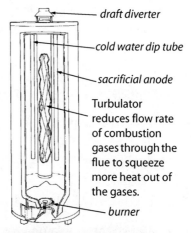

draft diverter

cold water dip tube

sacrificial anode

Turbulator reduces flow rate of combustion gases through the flue to squeeze more heat out of the gases.

burner

Standard gas water heater: These open combustion appliances are often troubled by spillage and backdrafting.

✓ A water heater must have a pressure-and-temperature relief valve and a safety discharge pipe. Install a relief valve and discharge pipe if none exists. The discharge pipe should terminate 6 inches above the floor or outside the dwelling as specified by local codes. The discharge pipe should be made of rigid metal pipe or approved high temperature plastic pipe.

✓ Adjust water temperature between 115° and 120°F with clients' approval, unless the client has a older automatic dishwasher without its own water-heating booster. In this case the maximum setting is 140°F.

✓Inspect faucets for hot-water leaks and repair leaks if found.

Electric Water–Heater Safety and Efficiency

Verify the following specifications for electric water heaters.

Gas

Electric

Setting hot-water temperature: Getting the temperature between 115 and 120°F can take a few adjustments and temperature measurements.

✓Electric water heaters should be serviced by a dedicated electrical circuit.

✓Specify replacement of damaged wiring and correct loose or improper wiring connections.

✓A replacement electric water heater should have an energy factor of at least 0.90 and be equipped with at least three inches of foam insulation.

8.3.1 External Insulation Blankets

✓Water heaters should be re-insulated to at least R-10 with an external insulation blanket, unless the water-heater label gives specific instructions not to insulate or the water heater is already insulated.

✓Water heater insulation must not obstruct draft diverter, pressure relief valve, thermostats, hi-limit switch, plumbing pipes, or element/thermostat access plates.

✓Keep insulation at least 2 inches away from the burner or gas valve.

✓Do not insulate the tops of gas- or oil-fired water heaters.

Electric Water-Heater Insulation

Observe the following specifications for insulating electric storage water heaters.

✓ Set both upper and lower thermostat to keep water at 120°F before insulating water heater.

✓ Insulation may cover the water heater's top if the insulation will not obstruct the pressure relief valve.

✓ Access plates should be marked on the insulation facing to locate heating elements and their thermostats.

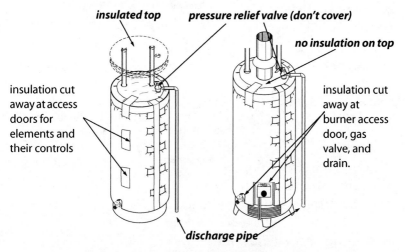

insulated top   pressure relief valve (don't cover)

no insulation on top

insulation cut away at access doors for elements and their controls

insulation cut away at burner access door, gas valve, and drain.

discharge pipe

Water heater insulation: Insulation should be installed carefully so it doesn't interfere with the burner, elements, draft diverter, or pressure relief valve.

8.3.2 Pipe Insulation

Observe the following specifications for insulating hot-water pipes.

✓ Insulate the first 6 feet of both hot- and cold-water pipes.

✓ Cover elbows, unions and other fittings to same thickness as pipe.

✓ Keep pipe insulation at least 6 inches away from flue pipe.

✓ Interior diameter of pipe insulation sleeve should match exterior diameter of pipe.

8.3.3 Water-Saving Shower Heads

Most families use more hot water in the shower than for any other use. A low-flow shower head reduces this consumption.

Replace high-flow shower heads with a water-saving shower head rated for a flow of 1.5 to 2.5 gallons per minute. Avoid installing the cheapest shower heads as they often provide a less satisfying shower and are prone to clogging.

Water-saving shower heads: The shower head on the right gives a laminar flow. The shower head on the left gives an atomized, steamy shower.

Use caution in removing the existing shower head from old, fragile plumbing systems.

Measuring Shower Flow Rate

You can determine flow rate by measuring the time it takes to fill a one-gallon plastic milk jug. If the jug fills in less than 20 seconds, your flow rate is more than 3 gallons per minute.

- ✓ Cut a large round hole in the top of the jug.

- ✓ Start the shower and set it to your judgement of a normal showering rate.

- ✓ Start a stopwatch at the same time you move the jug underneath the shower, capturing its entire flow.

- ✓ Note the number of seconds and divide 60 by that number to find gallons per minute.

8.4 INSTANTANEOUS DIRECT-FIRED WATER HEATERS

The switch to tankless water heaters is not a simple on because there are four generations of tankless water heater, which are described here.

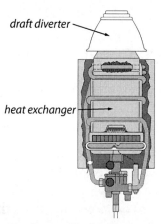

draft diverter

heat exchanger

1. Open-combustion tankless water heater with a draft diverter. This unit is more efficient than most storage water heaters, but not any safer in terms of draft.

2. Open-combustion fan-assisted tankless water heater. Slightly better than choice 1 in both efficiency and safety.

Obsolete tankless water heater: These older units may have a standing pilot, which negates some of the savings compared to storage water heaters.

3. Sealed-combustion fan-assisted tankless water heater (non-condensing). Solves the draft-safety problem with the same efficiency as choice 2.

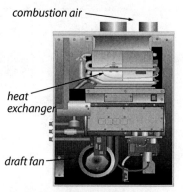

combustion air

heat exchanger

draft fan

4. Sealed-combustion, fan-assisted, condensing water heater. Solves the draft-safety problem and nearly optimizes efficiency. Consider using the same boiler for both space and water heating.

Sealed-combustion tankless water heater: These expensive water heaters have a tiny market share and save around one-third of energy used by the best storage water heaters.

5. Instantaneous direct-fired water heaters should be equipped with automatically modulating gas valves for satisfactory control of water temperature.

6. In regions with hard water, water should be treated or the instantaneous heater should be flushed with an acid solution at least every two years to avoid burnout of the unit's heat exchanger.

8.5 CIRCULATING DOMESTIC HOT WATER

In larger residential buildings, hot water is circulated to prevent long waits for hot water to arrive at fixtures and to prevent overheating in hot-water storage tanks.

Effective hot-water circulation requires a pipe to return the circulated hot water to the storage tank. In multifamily buildings, the return pipe returns water from the top of the hot-water mains or sometimes from the bottom of the mains. The hot-water circulation can be controlled by timers or occupant controls. Demand circulation systems use cold-water pipes as return lines for circulated water.

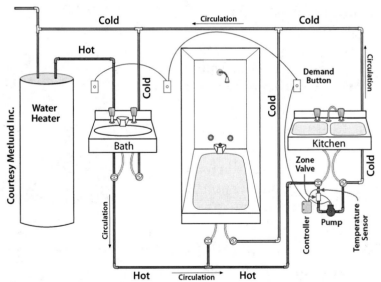

Demand hot-water circulation: The circulator is controlled by a temperature sensor, manual control, or occupancy sensor. Water returns to storage through the cold-water pipes.

8.6 Solar Water Heating

Solar water heating is a good option for many single-family and multifamily buildings. The success of solar water heating depends on the characteristics of the site and building including the following.

- The latitude and amount of solar radiation available at the site. This depends on climate, building orientation, and shading at the site.

- The potential to locate solar collectors near the main hot-water storage tank.

- The hot-water use patterns of the building and the potential for occupants to use solar hot water when it is available.

Designing a solar water-heating system to produce 100% of a building's needs isn't often wise because the system is so oversized in the summer that it may experience overheating problems.

Solar water-heating design and installation details are beyond the scope of this field guide. The following information is offered for general guidance only.

8.6.1 Solar Hot-Water System Design

Almost all North American climates require freeze protection for solar collectors. Two common designs provide reliable freeze protection using off-the-shelf components. These are the drain-back system and the closed-loop anti-freeze system. Both systems have five components in common with each other: solar collectors, heat exchangers, storage tank, circulators, and control system.

- The heat exchangers can have integral heat exchangers that surrounds the tank, heat exchangers immersed in the tank, of external heat exchangers.

- Storage tanks should be insulated to at least R-19.

- Pumps for drainback systems must overcome the static head of the system on startup, but pumps for closed-loop antifreeze systems must only overcome the system's friction head.

- Both systems should use an appropriate mix of propylene glycol and water for the heat-transfer fluid.

Drainback Systems

Drainback systems are more efficient, reliable, and durable than closed-loop antifreeze systems if installed properly. They have fewer components and the heat-transfer fluid won't stagnate because it drains back to a drainback tank when the high limit control deactivates the circulator.

- ✓The pump must be sized to overcome the static head between the tanks and the collectors.

- ✓The collector loop runs through a drainback tank where the heat-transfer fluid returns by gravity when the pump is deactivated.

- ✓The heat exchanger can be integral to the storage tank, immersed in the storage tank, immersed in the drainback tank, or external to both tanks.

- ✓All piping to and from the collectors must slope at least 10° and be a minimum of 3/4-inch diameter.

Closed-Loop Antifreeze Systems

Closed-loop antifreeze systems use a mixture of propylene glycol and water for the heat-transfer fluid.

- ✓The system must be pressurized according to the height of the collectors above the lowest point in the system.

- ✓A check valve must stop water from circulating through the collectors at night and wasting heat.

- ✓The system must be protected from overheating and damaging the glycol and collectors by the use of a photovoltaic-

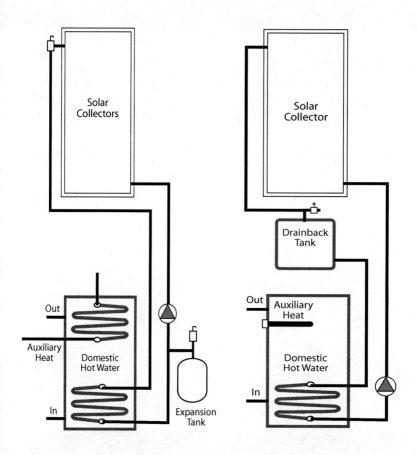

Closed loop antifreeze system: The most popular freeze protection.

Drainback system: Simple freeze protection with a minimum of valves and hardware.

powered circulator to keep water circulating or some other means.

✓The system must be equipped with an expansion tank and a 75 psi pressure-relief valve.

Water-Heating Energy Savings

LIST OF TABLES AND ILLUSTRATIONS

HVAC Energy Efficiency Service

Evaluating Combustion and Venting

Venting and Combustion Air

Heating System Installation

Evaluating Forced-Air System Performance

Cooling and Heat Pumps

Mechanical Ventilation

Water-Heating Energy Savings

INDEX